American Institute of Chemical Engineers

Topical Conference on Energy Processes 2007

Held at the 2007 AIChE Spring National Meeting

April 22-27, 2007
Houston, Texas, USA

Printed from e-media with permission by:

Curran Associates, Inc.
57 Morehouse Lane
Red Hook, NY 12571
www.proceedings.com

ISBN: 978-1-60423-358-2

Some format issues inherent in the e-media version may also appear in this print version.

ISBN: 978-1-60423-358-2

American Institute of Chemical Engineers
Proceedings
Three Park Avenue
New York, NY 10016-5991
Phone: 212-591-8100

www.aiche.org

American Institute of Chemical Engineers

Topical Conference on Energy Processes
2007

TABLE OF CONTENTS

Session 9 – Integrated Gasification Combined Cycle

Session 12 – Recent Advances in Fuel Cell and Battery Technology

Session 36 – Process Intesification for More Efficient Processes

ABSTRACT

Paper to be presented at the 2007 AIChE Spring National Meeting
Houston, Texas, on April 22-26, 2007

IGCC: Current Status and Future Potential

John Plunkett
Charles White
Salvatore Salerno
David Gray
Noblis
Glen Tomlinson
Consultant

This paper presents the results of a conceptual analysis of current and future IGCC configurations for the generation of electric power from bituminous coal. Two baseline IGCC facilities were established; one with no carbon dioxide capture and the other that captures and compresses carbon dioxide in preparation for geologic sequestration. The baseline plants were configured to produce about 600 MW of net power and utilize what is considered to be current technology. Single-stage, slurry fed, entrained flow coal gasification is used to produce the synthesis gas, and this is treated using conventional cold gas cleanup. In the non-capture configuration, the cleaned synthesis gas feeds two trains of FB-type gas turbines and steam is raised from the gas turbine exhaust to power a steam turbine in the combined cycle power block. In the carbon capture configuration, water gas shift converts most of the carbon monoxide to hydrogen, and the carbon dioxide that is produced is removed in the acid gas removal system. The gas turbine fuel is then predominantly hydrogen. The plant with carbon capture was designed to recover 95 percent of the carbon in the raw synthesis gas.

The U.S. Department of Energy aggressively sponsors R&D to improve the performance and reduce the cost of electric power generation from coal. Much of this R&D is directed towards future incorporation into IGCC configurations. Dry coal feed systems, warm gas cleaning, improved gasifier performance and availability, advanced gas turbines, membrane-based air separation, and solid oxide fuel cells are among the several areas of advanced R&D. Conceptual IGCC plants that incorporate these future improvements are being investigated. This paper reports on the predicted performance of advanced IGCC facilities for both carbon capture and non-capture, and compares them to those of the baseline plants. The value of this study is to assess the anticipated impact of DOE's R&D program on future IGCC process performance and cost.

Twenty (20) IGCC configurations that produce electric power from bituminous coal are being analyzed in this study. The impact of carbon capture is examined in the last eleven cases. All cases assume a carbon utilization of 98 percent. The cases have been grouped into two categories: the first includes cases without carbon capture (Cases 1 through 9) and the second includes the cases with carbon capture (Cases 10 through 20).

The method of analysis is to individually introduce emerging technologies that have the potential to improve overall system performance into the baseline IGCC plant in a step-wise manner, and then compare the results to observe the incremental improvements that result.

The specific technologies to improve the performance of the baseline IGCC plants that are examined in this study include:

- Dry feed gasifier, i.e., use of a Stamet pump to feed a single-stage, oxygen-blown gasifier
- Transport desulfurizer (TDS) and Direct Sulfur Reduction Process (DSRP) for acid-gas treatment and sulfur recovery
- Water gas shift reactor to produce CO_2 for carbon capture
- Warm gas cleanup for ammonia, chloride, and mercury removal
- Ion Transport Membrane (ITM) to supply oxygen to the gasifier
- Advanced syngas turbine (oxygen-fired and hydrogen-fueled)
- Solid oxide fuel cell

Technology improvements that have the potential to increase plant availability, such as longer life refractories, reduced maintenance time, and other reliability, availability and maintainability (RAM) improvements, are accounted for by increasing the capacity factor in several of the cases investigated. While this does not affect process performance, it does impact O&M costs. This paper will report on process performance. Near-term future work will evaluate the associated improvements in the cost of electricity (COE) and capital cost of all twenty cases. Longer-term future work will evaluate the effect of alternative gasifiers and coal feedstocks.

Table 1 summarizes the configurations of the nine (9) cases without carbon capture that are examined in this study. Figure 1 shows a block flow diagram of the process configuration for Case 1. Cases 2 through 9 involve variations to this baseline process.

Table 1
IGCC Technology Development Without Carbon Capture

Case	Description
1	Base Case / Slurry Feed Gasifier / 75% Capacity Factor (Current Technology)
2	Dry Feed Gasifier / 75% Capacity Factor
3	Dry Feed Gasifier / 85% Capacity Factor
4	Dry Feed Gasifier / 85% Capacity Factor / Transport Desulfurizer (TDS) and Direct Sulfur Recovery Process (DSRP)
5	Dry Feed Gasifier / 85% Capacity Factor / TDS and DSRP / Warm Gas Cleanup for Ammonia, Chloride, and Mercury
6	Dry Feed Gasifier / 85% Capacity Factor / Ion Transport Membrane (ITM) / TDS and DSRP / Warm Gas Cleanup
7	Dry Feed Gasifier / 85% Capacity Factor / ITM / TDS and DSRP / Warm Gas Cleanup / Advanced Syngas Turbine
8	Dry Feed Gasifier / 90% Capacity Factor / ITM / TDS and DSRP / Warm Gas Cleanup / Advanced Syngas Turbine
9	Dry Feed Gasifier / 90% Capacity Factor / ITM / TDS and DSRP / Warm Gas Cleanup / Solid Oxide Fuel Cell / Advanced Syngas Turbine

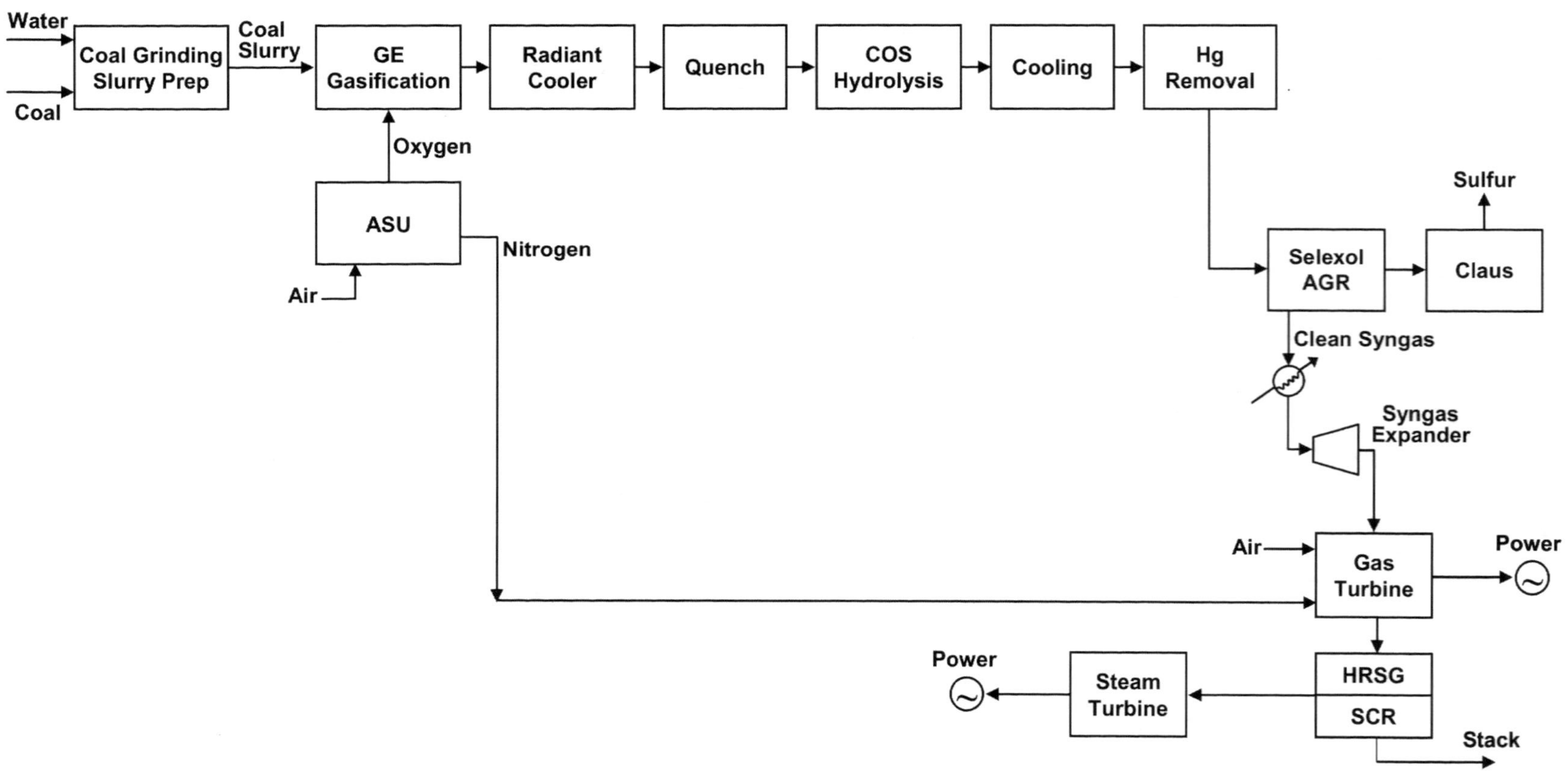

Figure 1

Case 1: Baseline IGCC Plant Without Carbon Capture

Table 2 summarizes the eleven (11) cases with carbon capture that are examined in this study. Figure 2 shows a block flow diagram of the process configuration for Case 10. Cases 11 through 20 involve variations of this carbon capture baseline process.

Table 2
IGCC Technology Development With Carbon Capture

Case	Description
10	Base Case Modified for Carbon Capture / Slurry Feed Gasifier / 75% Capacity Factor / Two-Stage Selexol for CO_2 Recovery (Current Technology)
11	Dry Feed Gasifier / 75% Capacity Factor / Two-Stage Selexol
12	Dry Feed Gasifier / 85% Capacity Factor / Two-Stage Selexol
13	Dry Feed Gasifier / 85% Capacity Factor / TDS and DSRP / Modified Selexol for CO_2 Recovery
14	Dry Feed Gasifier / 85% Capacity Factor / ITM / TDS and DSRP / Modified Selexol for CO_2 Recovery
15	Dry Feed Gasifier / 85% Capacity Factor / ITM / TDS and DSRP / Oxygen-Fired Gas Turbine
16	Dry Feed Gasifier / 85% Capacity Factor / ITM / TDS and DSRP / Oxygen-Fired Advanced Syngas Turbine
17	Dry Feed Gasifier / 90% Capacity Factor / ITM / TDS and DSRP / Oxygen-Fired Advanced Syngas Turbine
18	Dry Feed Gasifier / 85% Capacity Factor / ITM / Water Gas Shift / TDS and DSRP / Hydrogen-Fueled Advanced Syngas Turbine
19	Dry Feed Gasifier / 90% Capacity Factor / ITM / Water Gas Shift / TDS and DSRP / Hydrogen-Fueled Advanced Syngas Turbine
20	Dry Feed Gasifier / 85 % Capacity Factor / Solid Oxide Fuel Cell

Figures 3 and 4 summarize the predicted process performance for the nine cases without carbon capture and the eleven cases with carbon capture, respectively.

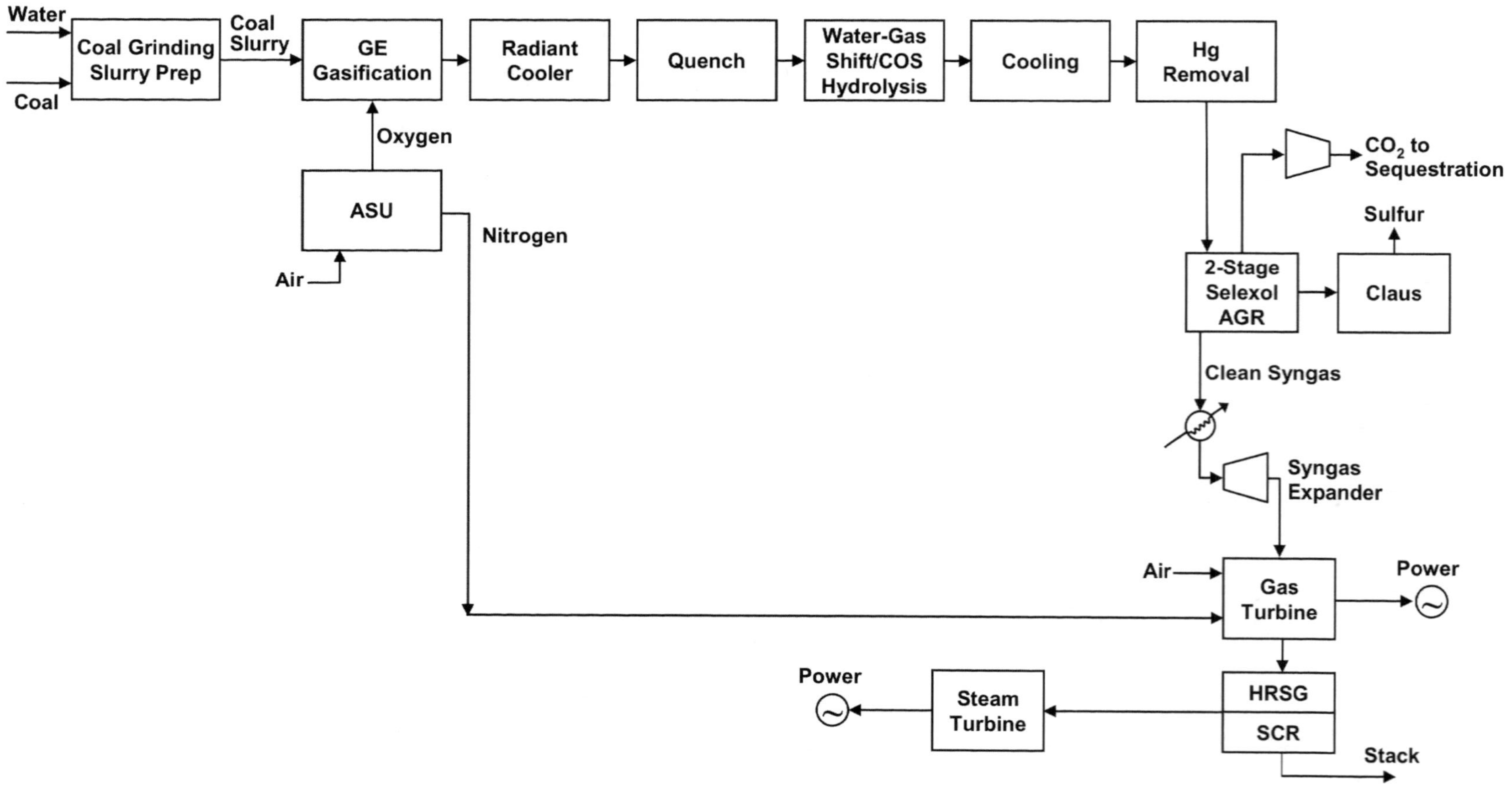

Figure 2

Case 10: Baseline IGCC Plant With Carbon Capture

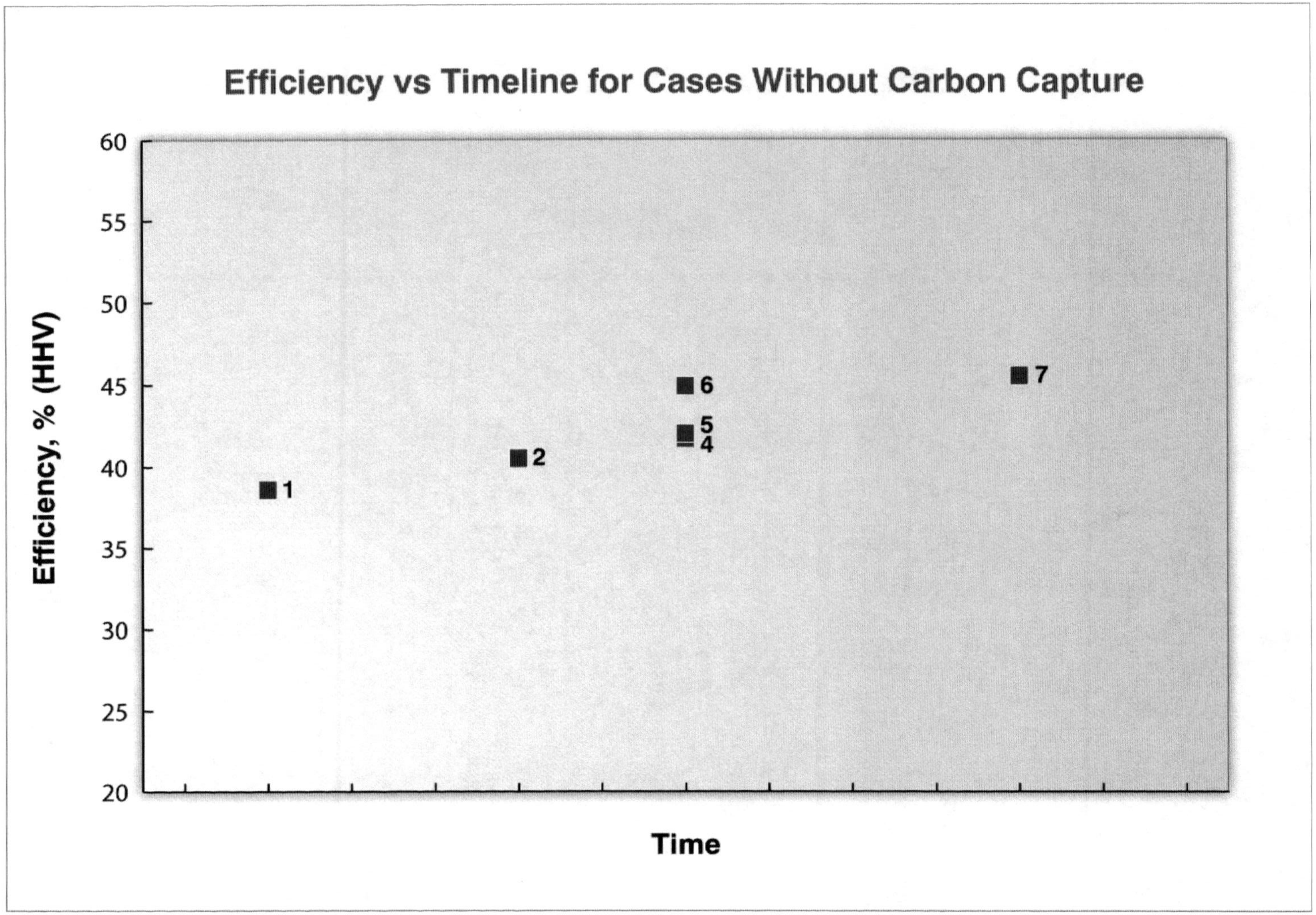

Efficiency vs Timeline for Cases Without Carbon Capture
Efficiency, % (HHV)
60
55
50
45
40
35
30
25
20
1
2
6
5
4
7
Time

Figure 3

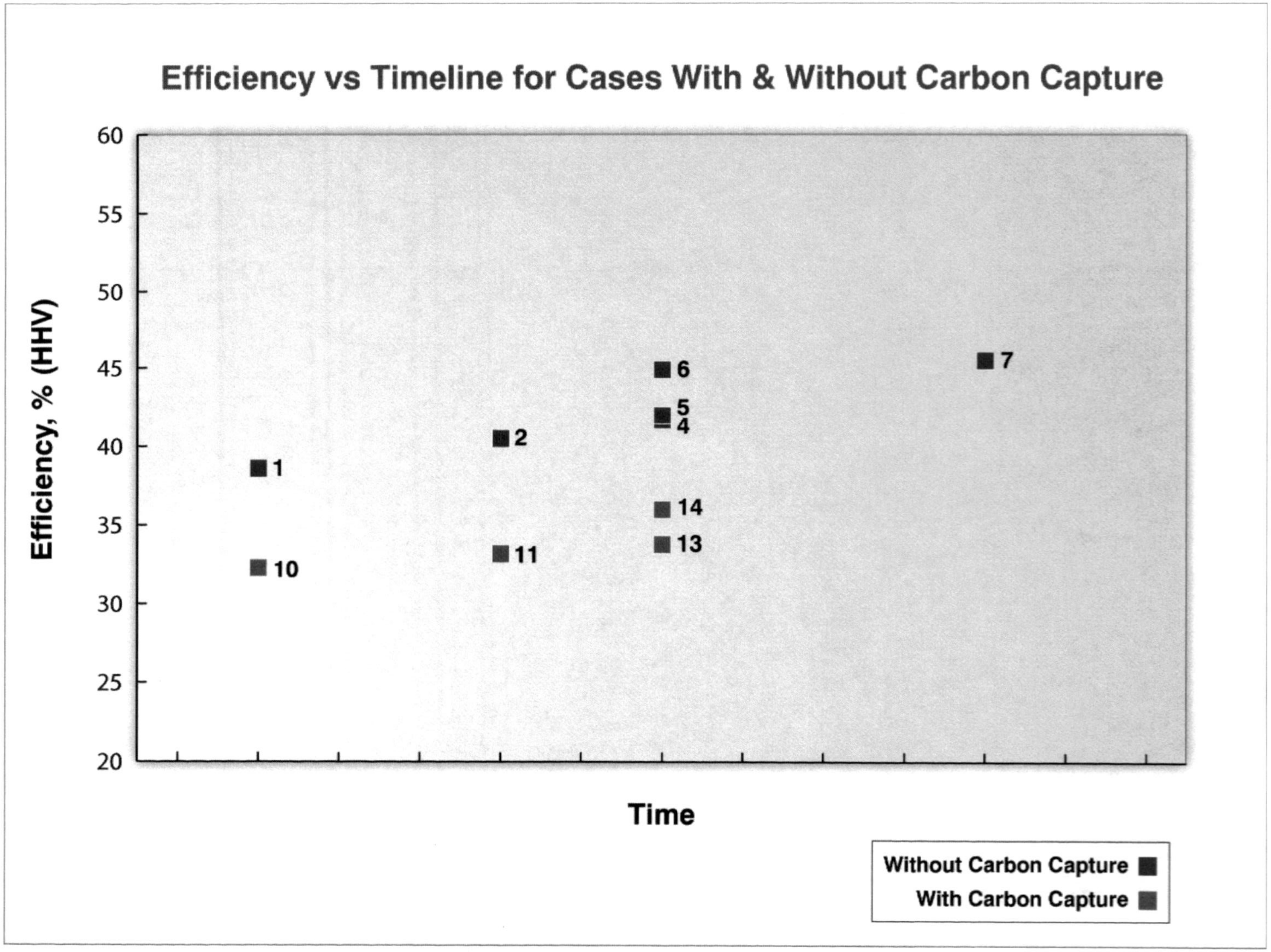
Efficiency vs Timeline for Cases With & Without Carbon Capture
Efficiency, % (HHV)
60
55
50
45
40
35
30
25
20
1
10
2
11
6
5
4
14
13
7
Time
Without Carbon Capture
With Carbon Capture

Figure 4

Comparing the baseline IGCC processes with and without carbon capture, thermal efficiency can be expected to decrease by about 16 percent, or by 6 percentage points. This is primarily due to CO_2 compressor power requirements and thermal energy used to generate steam for water gas shift. Improved thermal efficiency from IGCC processes both with and without carbon capture show potentially significant improvements resulting from emerging technologies. These are:

- Improved thermal efficiency from dry feed gasifier, saving the energy expended in vaporizing slurry water in the gasifier. Therefore, more of the coal's energy is converted to gaseous fuel.
- Improved thermal efficiency from warm gas cleanup. In cold gas cleanup processes, raw fuel gas must be cooled to below the dew point of water, thereby condensing water from the fuel gas and reducing its molar flowrate. The latent heat of vaporization is not recoverable; warm gas cleanup reduces that energy loss.
- Air separation using the ITM significantly improves IGCC performance. Integrated with the gas turbine, vitiated air produced by the ITM is at high pressure, thereby eliminating the need for the nitrogen compressor which is a significant user of plant auxiliary power.
- The advanced syngas turbine, with higher turbine firing temperature and therefore higher pressure ratio, improves the Carnot efficiency of the gas turbine. This turbine is also steam-cooled, which allows more air to flow through the initial turbine stages, thereby increasing power generated by the turbine.

At present, IGCC technology has about the same efficiency as supercritical pulverized coal plants but the future for IGCC power generation looks very promising if it is able to take advantage of advanced synthesis gas cleaning, advanced gas turbines, and eventually fuel cell topping cycles. IGCC also offers a viable solution for pre-combustion CO_2 recovery when carbon capture is mandated by future environmental regulations. From results observed during this study, emerging advanced technology R&D that DOE is sponsoring should enable increasingly efficient future power generation processes that will mitigate the performance and cost impacts of anticipated greenhouse gas emissions regulations.

Development of a Hydrogasification Process for Co-Production of Substitute Natural Gas (SNG), Biofuel and Electricity from Western Coal

Raymond Hobbs, Xiaolei Sun
Arizona Public Service, 400 N. 5th Street, Phoenix, AZ 85003

Abstract

Arizona Public Service (APS) Advanced Hydrogasification Project's (AHP) mission is to develop and demonstrate an engineering-scale coal hydrogasification-based process for co-production of Substitute Natural Gas (SNG) and electricity from Western coals, with zero CO_2 emission. The Phase I study focused on process alternatives selection, hydrogasification reactor configuration, process flow modeling, and bench scale experimental design. It is anticipated that 60% carbon in the coal can be hydrogasified. More than 50% methane yield is expected at the desired operating conditions of 1,600°F and 1,000 psi. Unconverted coal will be introduced into the oxy-burner to produce CO_2 and generate electricity. The CO_2 stream out of the oxy-burner is introduced to an algae farm. The algae farm can capture about 150 tons/acre-year carbon dioxide and produce 11,000 gallons biodiesel and 50 tons cattle meal annually per acre, based on GreenFuel Technology GEN3 performance tests. Bench scale lab testing is being initiated to mainly determine the energy and material balance around the gasifier and to study the hydrogasification kinetics. The effect of temperature, pressure, H_2/C ratio, H_2/O_2 ratio (for H_2/O_2 tip burner), coal residence time, and coal particle size will be investigated. The composition of the gas stream (CH_4, CO, CO_2, H_2S, H_2O, H_2, light HCs), the weight and composition of oil (light aromatics, tar), as well as ash (C, S) will be analyzed.

Keywords

Hydrogasification, Substitute Natural Gas (SNG), Western Coal, Algae, Biodiesel,

Background

The essence of the Arizona Public Service (APS) Advanced Hydrogasification Project (AHP) is to explore the hypothesis of converting coal to methane utilizing hydrogen as the "converter." This study is being done under a contract between DOE/NETL and APS, which began in April, 2006. The development of an advanced hydrogasification process resides within a broader national context of energy need, renewable energy, and a growing interest in CO_2 reduction. The AHP Project exists within that framework. Driven by the fact that in the near future the demand for natural gas will exceed North American supply and coupled with the world oil supply will be likely exceeded by real demand over the next 10 to 20 years, there is clearly a need for a broader range of energy options in the United States(Schlesinger 2005).

The AHP is designed to provide a source of sustainable, easily transportable energy – natural gas – starting within the next decade. It is also designed to fit within a timeframe when hydrogen will be available to the AHP from sources that are not available today, for example wind-generated hydrogen. Furthermore, work currently underway may result in hydrogen generated from off-peak electricity at power plants.

The AHP further anticipates greater value being placed on energy projects that do not release CO_2. By feeding CO_2 released from the process into an algae farm, CO_2 will be captured and consumed by the algae cells, where it is converted into hydrocarbon materials through photosynthesis. Oxygen will be released and valuable products (biodiesel, ethanol, and cattle meals) will be produced from the algae.

APS AHP Process

The APS AHP is a newly developed integrated process for the co-production of substitute natural gas (SNG), biofuel, and electrical power from western coals, with near-zero CO_2 emission. As shown in Figure 1, as-received Fruitland formation coal from northwest New Mexico goes to a hydrogasifier where it is treated with renewable hydrogen at a temperature of about 1,600°F at a pressure of 1,000 psia to produce methane-enriched gaseous products. The unconverted coal (primarily carbon and mineral matter) is withdrawn and sent to a low pressure oxygen combustion process (ash combustor), where CO_2, the main gaseous product, and slag is produced. A portion of this CO_2 stream is compressed and recycled to convey the coal into the hydrogasifier. The remainder is fed to the algae farm. The gaseous products from hydrogasifier are cleaned (fly ash, mercury, and sulfur are removed) and sent to a multi-bed methanation process to further convert CO and CO_2 into CH_4. The raw SNG leaving the methanation process is cooled, flashed to remove water, dehydrated to remove residual water, and passed through a membrane process to remove extra hydrogen. The finished SNG is compressed and injected into the El Paso natural gas pipeline system.

The AHP process was developed to utilize the US's abundant coal supply and renewable hydrogen to make the SNG (CH_4). Coal provides the carbon for the process. Wind energy can be used to power an electrolyzer that splits water into oxygen and hydrogen needed for the hydro-gasifier. Algae, fed on carbon dioxide emissions, is dried and used either as an additional fuel source or as a feedstock for biodiesel and cattle meal production. The role for each of these concepts, or the utilization of alternative concepts or a combination thereof, is under continuing evaluation.

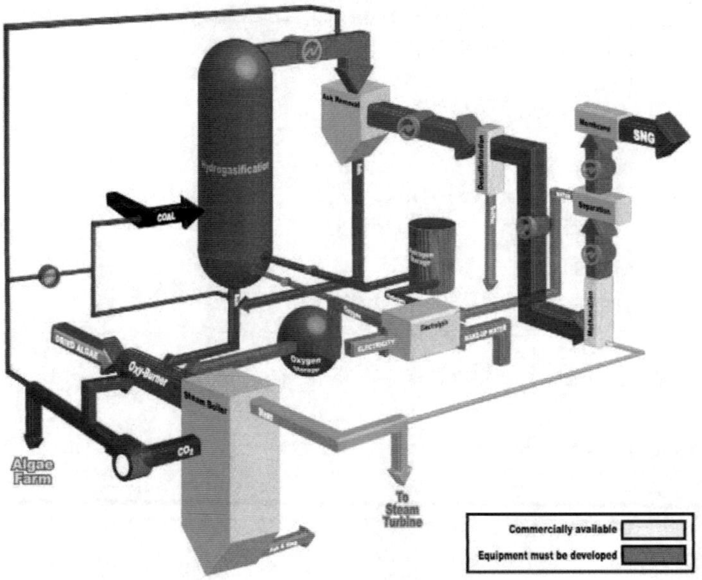

Figure 1 APS Advance Hydrogasification Process

The resulting system could be placed at a mine mouth location in the Four Corners area, far from load centers, with decent wind energy resources (or other source of hydrogen). The SNG can be transported by natural gas pipeline to highly efficient, clean natural gas combined cycle (NGCC) power plants. Efficiency will be gained by eliminating transmission line losses and by providing a reliable source of natural gas that is so critical for meeting peak load demands in the West. Electric power produced in the AHP process and the SNG used in NGCC power plants meets the stringent new California long-term power purchase performance standard for greenhouse gas emissions. The assessment from the Western Governors' Association indicates that the Governors supported advanced coal technologies, such as hydrogasification, as an important resource for the future. Recognition of the importance of diverse and reliable sources of electric power that reduce emissions is clearly growing rapidly in the western US.

The performance targets for the AHP include:

- A process efficiency greater than 50%;
- Production of SNG at a cost of \$5 - \$10/MMBtu;
- Capture of more than 90% of emission from a coal-fired generating unit;
- Reduction in water use by least 50% below that of a comparable gasification or synthesis gas methanation process plant;
- Ability to accept hydrogen as a supplemental source of energy;
- Use of low-rank western coal;
- Co-production of electricity;
- Evaluation of an algae process for CO_2 recycle and drafting of a test protocol for the algae process;

Process Modeling

An AspenPlus™ process simulation model has been developed for this AHP process. From 1,000 tons/day (37,799.4 kg/hr) of dry Fruitland coal, simulation results show the process produces 25,510 kg/hr of SNG with an assumed 70% carbon conversion (Feldmann 1973; Asaoka 1999; Noguchi 2000). The process requires about 7,000 kg/hr hydrogen and 27,500 kg/hr oxygen. Besides the SNG, the process also produces 261 kg/hr of sulfur and exports about 41 MWe of electric power. The CO_2 generation rate is about 400 ton per day, which will be fed to an algae farm, which will produce about 60 MWe dry algae.

The design basis for the AspenPlus™ simulation model is presented below.

End Product Required:	Substitute Natural Gas (SNG) for export to pipeline
Gasification Processes:	Hydrogasification and Ash Combustion
Coal Feed Rate:	1,000 stpd (dry basis) or 1,078 stpd as-received (7.27% moisture)
Oxygen Purity:	100% oxygen at 1,025 psia and 275°F (135°C)
Hydrogen Purity:	100% hydrogen at 1,025 psia and 275°F (135°C)
Gasification Pressure:	1,000 psia
Gasification Temperature:	1607°F hydrogasifier exit
Sulfur Removal:	Crystasulf process
Methanation:	Multistage methanation process with high conversion of both CO and CO_2 to methane
Product Treating:	Drying and compression (to 875 psia)

The process has an overall thermal efficiency of 94% based on the higher heating values of the feeds and products. When the electrical energy that is required to produce the externally generated hydrogen and oxygen is considered, the overall thermal efficiency drops to 78% on a thermal HHV basis, but there are 51,271 kg/hr of pure oxygen available for other uses or sale.

The following table (Table 1) (Nexant 2007) shows the overall battery limit flows for the process.

Table 1 Overall Mass Balance of APS AHP from Aspen Simulation

Inputs	Kg/hr	Lb/hr	stpd
As-received Fruitland Coal	40,762.8	89,867	1,078.4
Hydrogen	6,989	15,407	
Oxygen	27,096	59,736	
Total Mass Input	74,847	165,010	
Outputs	Kg/hr	Lb/hr	MMscfd
SNG	25,522	56,267	31.65
Flue Gas to Algae Farm	17,738	39,105	8.98
Sulfur	261	575	
Hydrogen Purge	100	221	
Slag	8,059	17,767	
Waste Water	23,151	51,039	
Crystasulf Waste	16	36	
Total Mass Output	74,847	165,010	
			MWe
Estimated Export Power			41.2

Based on the Aspen modeling, external heat is required to maintain the thermal neutral operation of the hydrogasifier. To get this additional heat, oxygen is introduced to burn a small portion of hydrogen in a H_2/O_2 pre-burner (Friedman 1979; Asaoka 1999; Noguchi 2000). By this means, the hydrogen stream is heated to a high temperature before it contacts the coal to initiate the hydrogasification reaction. A manual trial and error procedure was used to calculate the oxygen feed rate going into the hydrogasifier. The oxygen feed rate was set to maintain the desired hydrogasifier exit temperature of 875°C (1607°F). The simulation result shows that a H_2/O_2 mole ratio of 30:1 is required in the pre-burner. This combustion generates about 5,600 kg/hr water vapor.

The pure oxygen flow rate to the ash combustor automatically was set so that there was 10% excess oxygen leaving the ash combustor. The excess hydrogen was removed from the SNG product and recycled back into hydrogasifier.

Using the same simulation model, a series of sensitivity cases were also investigated to study the effect of reactor temperature, carbon conversion, hydrgen/coal ratio, addition of CO_2 and H_2O et al.

Estimated SNG Price

SNG price has been estimated for the proposed AHP process. The price is heavily dependent on the method of hydrogen supply. As shown in Figure 2, in current APS AHP process there are three main strategies considered for hydrogen supply. Path I was through electrolysis powered by wind electricity. This pathway provides 100% renewable hydrogen. However, due to relatively high cost of wind plant and electrolysis plant, the estimated hydrogen price is somewhere between $2-5/kg(Leighty 2007). Hydrogen storage needs to be considered since wind electricity does not have a firm delivery. In Path II, the electricity required by electrolyzer will be provided by APS internal grid electricity, which tends to be the off-peak electricity. It is likely that the hydrogen price will drop to between \$1 – 2 /kg. Path I and Path II can be combined since they all use electrolysis process to produce hydrogen. Path III is to supply hydrogen through gasification method. In this "Coal to Hydrogen" process, the coal is first gasified with oxygen and steam to produce synthesis gas consisting primarily of CO and H_2, with some CO_2, sulfur, particulates and trace elements. In "Hydrogen from Coal Program" report by US DOE (Longanbach 2002; 2006), three different further processes are described for hydrogen and electricity generation from synthesis gas. The hydrogen price is claimed to be anywhere between \$0.4 – 1.1/kg.

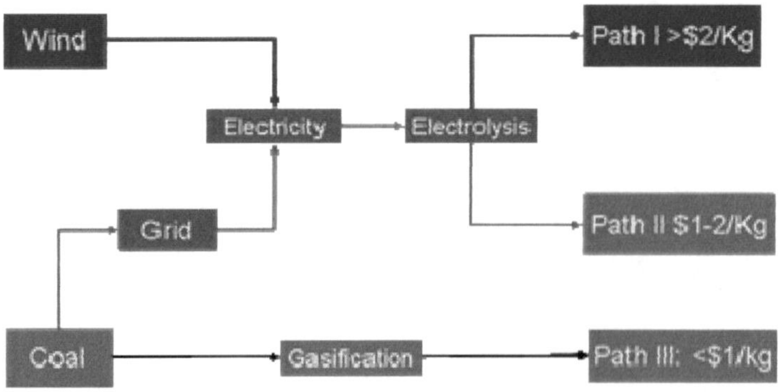

Figure 2 Hydrogen Supply Source

Combining with estimated capital, O&M cost of hydrogasification process and electrolysis process, Figure 3 gives the estimated SNG price. Profits from co-generated electricity, biodiesel and cattle meals are also included in the calculation. As shown in the figure, the current estimated SNG price falls between \$4 - 10/MMBtu. Using 100% wind electricity does not seem practical at the moment due to the scale of the project and the current availability/price of wind electricity.

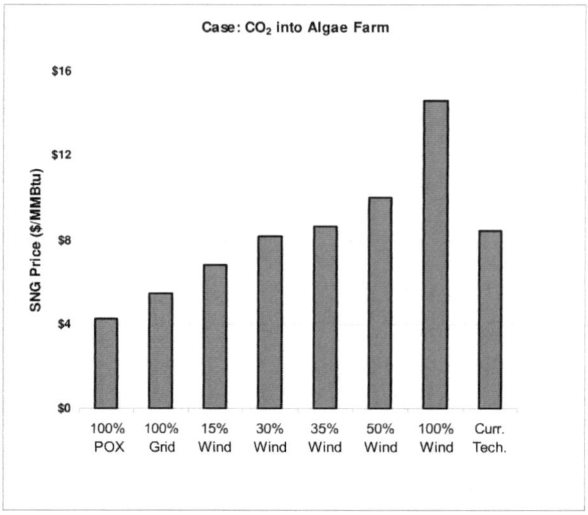

Figure 3 Projected SNG Price

Coal Hydrogasification

Fruitland Coal

The project analyzed Fruitland coal from field in the Northwest New Mexico. Table 2 summarizes the coal physical and chemical characterization. This western coal has relatively low moisture and high ash content. On a dry basis, it has a heating value of 10,710 Btu/lb.

Coal High Pressure TGA (HPTGA) Test

Three HPTGA tests have been successfully completed by Gas Technology Institute, Chicago, IL, to evaluate the gasification reactivity of the devolatilized Fruitland coal sample in the 1500-1700°F temperature range, at 500 psi pressure using a 50 vol% H_2O – 50% N_2 gas mixture. One additional test was also carried out at 1700°F, but at the higher pressure of 700 psig. HPTGA test results indicate the devolatilized Fruitland coal sample is highly reactive under these regular gasification conditions. Consistent with expectations, reactivity improved with increasing temperature. Reactivity was reduced at the increased pressure. It took 90 minutes to achieve ~100% base carbon conversion at 1500°F (and 500 psig). Similar high conversions were achieved during 30 minutes at 1600°F and during only 15 minutes at 1700°F. At 1700°F, reactivity at 700 psi was lower than that at 500 psig, requiring about 25 minutes to achieve complete conversion compared to 15 minutes at the lower pressure.

Table 2 Properties of Fruitland Coal

Proximate Analysis	As Received	Dry Basis	
Moisture, %	7.27	---	
Volatile Matter, %	35.00	37.74	
Ash (950°C), %	20.07	21.32	
Fixed Carbon, % (by difference)	37.66	40.94	
Ultimate Analysis	Dry Basis	Fluorine, Chlorine, and Trace Elements (µg/g)	
Ash (750°C), %	21.32	Fluorine	87
Carbon, %	61.92	Chlorine	99
Hydrogen, %	4.61	Arsenic	< 200
Nitrogen, %	1.28	Cadmium	< 6
Sulfur, %	0.69	Lead	< 200
Oxygen, % (by difference)	10.18	Mercury	0.055

The hydrogasification performance of Fruitland coal was tested using the same technology. A raw coal sample (100 milligrams, -20+50 mesh) was used. The testing conditions were 1600°F, 800 psig, and 90% H_2-10% N_2 mixture. With a heating rate of 1000°C/min, coal sample reached the desired test temperature within a few seconds. The test took 96 minutes until no further weight change was measured. Base carbon conversion was estimated as 84.8%. This lower conversion suggests that the coal is not as reactive under hydrogasification conditions as regular gasification conditions. The process may tend to be a kinetic controlled process. Even lower conversion should be expected during the bench/engineer scale operation because of the limitation on the achievable residence time.

Figure 4 shows Scanning Electron Microscopy (SEM) images of fresh coal, char from HPTGA test and fly ash from power plant. SEM confirmed cavity structure in the char sample which indicates entrapped volatiles burst during the testing. Fly ash is essentially fine particles (<10 μm). Energy Dispersive Spectroscopy (EDS) indicates that fly ash is mainly composed of aluminum. However, both the fresh coal and char samples from HPTGA are mainly composed of carbon and oxygen, with trace amounts of aluminum.

Figure 4 SEM of Fresh Coal, Char Sample from HPTGA, and Fry Ash from Power Plant

Bench Scale Testing

The proposed AHP process combines off-the-shelf technology with the advanced technologies in the hydrogasifier and Algae farm. APS is building a bench-scale hydrogasifier. The objective for bench-scale testing will be to experimentally determine and/or verify key assumptions for the APS AHP process. Data obtained from bench-scale testing will be used to study the kinetics of western coal hydrogasification. The knowledge will also be used to update ASPEN process modeling, support commercial scale hydrogasification reactor CFD modeling and guide the next phase experimental testing. Key test parameters are temperature and pressure, coal particle size, residence time, hydrogen to carbon ratio and coal particle size. The bench-scale test reactor must be capable of varying temperature and pressure, coal particle size, residence time and hydrogen to carbon ratio to experimentally determine optimum conditions for methane production. The current bench reactor design is 1.6" i.d. and 12 feet long. It can provide a test range of 5 - 15 lb/hr coal feed rate, 300 -900 SCFT hydrogen feed rate, and up to 1900°F temperature and 1200 psig pressure. The system also has the capability to test the effect of H_2O and CO_2 additions. For simplicity, the H_2/O_2 pre-burner design is not introduced into this bench reactor. H_2 is heated by a pre-heater and the whole reactor is heated by external heaters.

Performance of Algae Farm

Working with Green Field Technology (GFT), a demonstration pilot algae system (GEN3) was built at the APS Red Hawk natural gas-fired combined cycle (NGCC) power plant. Actual results from the Red Hawk GEN3 system demonstrated an algae growth rate of 57 grams per square meter-day. This growth rate extrapolated to 150 tons/acre-year of carbon capture which could yield about 11,000 gallons of biofuel annually per acre. Economic modeling estimates the cost of algae-based transportation fuel from this process to be competitive with petroleum based fuels, when systems are built to moderate scale. In addition, the adsorption of carbon dioxide is 70 times the rate per unit of land compared to conventional agricultural operations. The corresponding growth rates result in yields that produce 20-70 times more biofuel than other conventional crops including corn, canola, and jatroha trees, without using limited resources as fresh water and fertile land (shown in Table 3).

Table 3 Diagram Biomass Performance

Crop	CO_2 Capture (tons/acre/yr)	Biodiesel Yield (gals/acre/yr)	Ethanol Yield (gals/acre/yr)
GFT Algae process	150	8,000	3,000
Corn (Source2)	<6	18	328
Soybean (Source2)	<2	48	n/a
Canola (Source2)	<2	127	n/a
Jatropha (Source2)	-	194	n/a

An Engineering Scale GEN5 System has been constructed at Red Hawk and will undergo performance testing. The GEN5 system is expected to capture higher amounts of CO_2 per acre at lower costs and at higher efficiency than the GEN3 system. Higher capture rates will result in greater amounts of algae per acre.

Conclusions

One of the key challenges in the coming decade is to produce electricity in a more sustainable manner without the production of large quantities of carbon dioxide emissions. This project is aimed at achieving this goal by developing and demonstrating an engineering-scale coal hydrogasification-based process using western coal and renewable hydrogen source. Sustainable SNG will be supplied and complete carbon recycle will be accomplished using an algae farm.

The proposed APS AHP process combines a hydrogasification reactor to produce methane tied to generating unit to burn excess char from the gasification (co-production of electricity) and a back end clean up utilizing algae to reformat the CO_2. The algae process essentially allows for the recycling of carbon back into the process, which makes the designed hydrogasification process is "zero emission". The Algae produced could also be feedstock to produce Biodiesel and cattle meal. There are multiple variables in utilizing renewable sources for hydrogen production and for applying algae to capture CO_2 from the SNG process or electric generation facilities. The AHP process will have overall energy efficiency above 78%. An Algae farm with a GEN3 system can capture 150 tons/acre-year of carbon dioxide and can yield about 11,000 gallons of biofuels (biodiesel and ethanol) per acre year.

Acknowledgements

The project is funded by DOE/NETL under grant DE-FC26-06NT42759. The authors would like to thank Robert Page, John Boyle, Fuyuki Noguchi (*Arizona Public Service, Phoenix, AZ*), Donald Karner (*eTec, Phoenix, AZ*), Elaine Everitt, Chris Guenther (*NETL, Morgantown, WV*), John Ruby, Sheldon Kramer (*Nexant, San Francisco, CA)*, David Stauffer, Jay White *(WorleyParsons, Houston, TX),* Alan Weimer *(University of Colorado, Boulder, CO)*, Bruce Bryan *(Gas Technology Institute , Chicago, IL*) for their joint effort.

References

(2006). Hydrogen from Coal Multi-Year RD&D Plan, US DOE.

Asaoka, Y., Azuma, T., Gray, H., Noguchi, F., Maruyama, H. (1999). Development of Coal Hydrogasification Technology 1. Hydrogasification and Simulation in the ARCH Gasifier. 16th Annual International Pittsburgh Coal Conference.

Feldmann, H. F., and Yavorsky, P.M. (1973). The Hydrane Process. 5th AGA/OCR Synthetic Pipeline Gas Symposium, Chicago, Illinois.

Friedman, J. (1979). Development of a Single-stage, Entrained-flow, Short-residence-time Hydrogasifier. Canoga Park, CA, Rockwell International.

Leighty, W. C. (2007). Confidential: Renewable-source Hydrogen Supply for the AHP* Plant.

Longanbach, J. R. (2002). Hydrogen Production Facilities Plant Performance and Cost Comparisons, NETL and CTC.

Nexant (2007). "Internal Report to APS."

Noguchi, F. (2000). Coal hydrogasification process for the production of SNG and chemicals. Annu. Int. Pittsburgh Coal Conf.

Schlesinger, J. (2005). "Thinking Seriously About Energy and Oil's Future." The National Interest **60**: 19-24.

Syncrude and Syncoal Production by Mild-Temperature Pyrolysis Processing of Low-Rank Coals

I

Ebbe R. Skov, P.E., ConvertCoal Inc., Mission Viejo, CA, Hetagon@aol.com
Dennis C. England, Ph.D., P.E., E2 Environmental, Inc., Irvine, CA
James C. Henneforth, P.E., Pacific Group Electric Power, LLC, Coto de Caza, CA
Franklin G. Rinker, P.E., ConvertCoal Inc., Naples, FL

Prepared for Presentation at AIChE Spring National Meeting, Houston, Texas, April 22 – 26, 2007
AIChE SESSION – TB004 - #9c, 23 April 2007

ABSTRACT The novel coal-to-liquid (CTL) pyrolysis process developed by ConvertCoal, Inc., converts bituminous, sub-bituminous and lignite low-rank coals (LRC) into low-emission clean-coal power boiler fuel and syncrude after hydrotreating the recovered coal-tar oil. The processed clean-coal product can meet the U.S. latest Clean Air Interstate Regulations taking effect in 2010 with little or no additional emissions mitigation, and fuel efficiency also is improved resulting in decreased green-house gas, or GHG, (carbon dioxide) emission. The simultaneous production of clean coal and syncrude oil provides an economic basis for this new CTL process. The syncrude results from on-site catalytic hydrotreating of recovered coal-tar-oil in 15 - 20w% yields on a water-and-ash-free basis, indicating syncrude yields of 0.75 – 0.95-bbl/ton of feed-coal. ConvertCoal, Inc., has recently completed the design and patent applications for a modular mild-temperature pyrolysis plant for processing 10,000-t/d Wyoming PRB coal to produce 1150-t/d coal-tar-oil and 5000-t/d low-emission coal-char fuel matching a 500-MW power plant. The hydrogenated coal-tar-oil yields 8000-bbl/d syncrude with suitable properties for petroleum oil refining. A new "virtual oilfield resource" producing 200,000-bbl/d with predictable oil quality and defined recovery profile therefore would result from having in operation two dozen such projects corresponding to 12,000-MW generating capacity and 84-million-t/y LRC (less than 15% of annual LRC mining capacity). The capital intensiveness of the project is similar to oilfield and power projects on a barrel-per-day of oil or ton-per-day of coal basis. Current engineering studies indicate that the process can be economically attractive at current energy price levels. It therefore could be the basis for converting the large U.S. LRC resources into equivalent oil reserves. In addition, due to the heating value increase in the processed coal-char, the process enables IGCC projects to use less costly LRC on a par with high-rank Eastern high-sulfur coals. Increased interest in the ConvertCoal process therefore can be anticipated based on the need for a reliable supply of low-emission clean coal, a strong desire for the development of domestic syncrude oil resources, and the existence of enormous LRC reserves in the Western U.S. and other coal rich countries. //.

INTRODUCTION

The demand for crude oil worldwide is expanding at a 2% annual rate or by 1.8-million-daily barrels, resulting in recent price increases. Coincidentally, coal-power generating plants are required to reduce emissions of sulfur dioxide, nitrogen oxides and mercury. For these reasons, a coal-conversion process with the capability of expanding the domestic crude oil supply and reducing the coal-power emissions will be of considerable interest at this time. Such a coal-conversion process now is available and cost effective based on using low-rank-coals (LRC) that comprise the most abundant coal resources in the U.S. and worldwide.

The coal-to-liquids (CTL) process developed by ConvertCoal, Inc., (CCI) is designed to convert low-rank-coals (LRC) that typically contain 30 – 35w% "volatile matter" into two products. The first is coal-tar-oil that can be hydrotreated to produce syncrude oil for petroleum refining. The second is a low-emissions coal-char fuel suitable for PC-power generation plants. The CCI proprietary SynCrude-Syncoal™ (SC2) process is particularly well suited for LRCs. The coal drying and mild-temperature pyrolysis process is described in the following, including a summary of the process development history, process design, selection of feedstock, and the products characteristics and applications. Additional information about the coal-tar oil recovery and conversion to syncrude oil is presented in this conference as separate paper (ref.-1).

MILD TEMPERATURE PYROLYSIS PROCESS HISTORY

Several mild-temperature pyrolysis (MTP) processes were developed during the past century, and successful examples of commercial processes included the COALITE in Britain, Ruhr-Chemie in Germany and SASOL in South Africa. MTP processes operate at up to 650°C (1200°F) and relatively short reaction times compared with metallurgical coking that operates at up to 1000°C (1830°F) and 6 – 8-hour cycle times. The composition of the feed-coal also is different, and the coal-tar-oil (CTO) from MTP therefore is different from coking-oven tar-oils. In the past, the market for coal-tar chemicals has been supplied for the most part from high-temperature metallurgical coking plants, and generally MTP has not been economically viable at prevailing energy price levels in direct competition with petrochemical products.

Between 1977 – 1997, considerable work was performed to advance and test several new MTP technologies in the U.S. with DOE and industrial support, and notable advancements were made by several companies using processes operating in the 600°C (1100°F) temperature range in pilot demonstration plants (Table – 1). In 1958, the U.S. Bureau of Mines investigated pyrolysis of LRC at low, medium and high temperatures on a laboratory scale (ref –2).

For example, a joint venture between Shell Mining Corp and SGI International, ENCOAL Corp., designed and operated a 1000-t-coal/d demonstration project with DOE support between 1992 – 1998 near Gillette, Wyoming producing low-emission coal-char-fuel that was successfully tested in several power generation plants. The plant proved the technical feasibility of using LRC in the MTP process, established correlation between feedstock composition, operating conditions, recovered coal-tar-oil quality and yields, and also identified several important design features needing improvement. Moreover, the commercial success over many years of the COALITE Corp., as producer of coal-tar chemical intermediate products from British coals has demonstrated the viability of the MTP process. The proprietary CCI process presented here was developed over the past several years using process engineering modeling based on data acquired from the DOE programs, design optimization and incorporation of novel process design features for which patents have been applied.

TABLE – 1
EXAMPLES OF MILD-TEMPERATURE PYROLYSIS PROCESS DEVELOPMENT

COALITE Corporation, British producer of coal-tar chemical intermediates

Allis Chalmers, MTP processing of coals in rotary kilns

TOSCO, indirectly heated solid-contact heat transfer pyrolysis of LRC, 1978

ERC/AMAX Circulating Fluidized Bed Reactor R&D project: Development of an Advanced, Continuous Mild Gasification Process for the Production of Co-Products; Final Technical Progress Report for Period Feb 1988 - Mar 1989; Dec 1989; US DOE DE-AC21-87MC24267

DOE-MC, Development of an Advanced, Continuous Mild Gasification Process for the Production of Co-Products. Final Report Sept 1987 - 1996; DOE/MC/124116 - 5297; US DOE Contract No.: DE-AC21-87MC24116

Western SynCoal LLC, SGWR Report, Aug 2003, US DOE Grant DE-FG26-00NT40965; coal drying and pyrolysis of LRC, Montana; www.osti.gov/bridge/

SGI/SMC–ENCOAL Corporation, 1000-t/d demonstration plant, Wyoming, 1992 – 1998; ENCOAL Project Final Report DOE/MC/27339-5798, US DOE Sept 1997

FMC Corporation, multiple fluidized bed drying and pyrolysis reactors, Char Oil Energy Development (COED), FE-1212-T-9 Report September 1975.

CCI SYNCRUDE-SYNCOAL PROCESS DESIGN

The CCI process includes a series of processing steps that integrate conventional unit operations and equipment into the continuous-flow design. The process design has been optimized in several important ways to improve the product yield, energy efficiency and operability while also reducing the capital cost. The three core processes sections are coal drying, pyrolysis reaction and coal-tar-oil (CTO) recovery. The supporting downstream processing sections include water and pyrolysis-gas cleanup, CTO fractionation and hydrotreating to syncrude oil, hydrogen production and cogeneration of steam and electricity. A process summary block flow diagram is included at the end of this paper, followed by a flow diagram for the drying, pyrolysis and CCF processing. To better illustrate the narrative with numeric examples, we have used an average sub-bituminous low-rank coal (LRC) from the Buckskin mine located near Gillette, WY. The coal contains approximately 30w% moisture, 32w% "volatile matter", 5w% ash, 3w% pyrolysis-water, 0.4 – 0.8w% sulfur and 180-ppb mercury.

COAL DRYING As a first step, feed-coal is crushed and screened to "minus 6-mm" and then conveyed into the coal-drying section where the moisture is reduced first to 8w% at 100°C (212°F) and further to 3w% at 270°C (520°F). Small amounts of gasses volatilize with the water, including some CO, CO_2, NH_3, CH_4 and H_2S. Up to 35% of the sulfur and nitrogen content in the coal also may be removed in this process step depending on the feed-coal composition, and 85 – 95% of the volatile mercury compounds will be removed. Several configurations of equipment in commercial use can be selected for this process step. Notable process design parameters include the amount of water removal, means of heat transfer, process gas velocity, degree of coal comminution, mercury removal and recovery, and energy optimization.

PYROLYSIS REACTION After drying, the dried coal proceeds to the pyrolysis reactor where it is heated to 550°C (1020°F) and approximately two thirds of the "volatile material" in the feed-coal is removed as condensable coal-tar-oil and non-condensable fuel-gas. The balance is left in the residual coal-char fuel (CCF) to ensure sufficient volatility for good ignition and flame stability. The pyrolysis process removes the remainder of the volatile mercury compounds from the CCF and more than half of the organic sulfur and nitrogen compounds remaining after drying.

A combination of direct and indirect heating will be used in the pyrolysis unit as the coal is contacted with hot inert sweeping-gas in counter- or cross-current flow for heating and mass transfer of the volatile coal-tar-oil. The recovered oil quality and yield depend on both the feed-coal composition and pyrolysis reactor time/temperature profile, and these operating parameters predominantly determine the overall process economics. Various equipment configurations are possible for conducting the pyrolysis reaction, including horizontal travelling grates, rotating grates, or rotary kilns commonly used in minerals and petroleum-coke calcining operations. In particular, the process design takes into consideration the coal softening temperature range, temperature profile and residence time, gas phase velocity, CTO yield and gas-phase concentration, coal comminution and coal fines production, and the efficiency of heat transfer and heat recovery.

COAL-CHAR COOLING AND PYRITE REMOVAL During the pyrolysis at 550°C (1020°F) iron-pyrite [FeS_2] will in part have converted to paramagnetic pyrrhotite [FeS_x] that now can be separated. The CCF product exiting the pyrolysis reactor is cooled in a separate heat-exchange unit to 50°C (120°F) and further crushed and screened to minus 1-mm. It is then processed in a "magnetic material-separation" unit for removal of pyrrhotite, reducing the ash and residual sulfur content of the CCF product. The finished product is finally conveyed to an intermediate storage silo, from where it can go into the coal-char stabilization unit or directly into the adjoining PC-power plant pulverization unit. The equipment selected for this process section is found in commercial operation in various industrial applications.

COAL-CHAR STABILIZATION For purposes of longer time storage and transportation without using inert gas blanketing, the bone-dry CCF product will require "stabilization" against self-ignition in the open air, because heat generated from adsorption of atmospheric oxygen and humidity can cause spontaneous combustion. Stabilization is accomplished in a separate processing unit using carefully controlled amounts of air oxidation and humidification, adding back a total of 3 – 5w% oxygen to the CCF product. This process step has been demonstrated in several commercial capacity plants.

COAL-FINES HANDLING Processing of LRC feedstock inevitably will result in comminution of some amount of the coal into coal-fines, and the amount depends on the specific coal, the equipment selection and actual processing conditions. The production of coal-fines, if not excessive, is advantageous for the adjacent PC-power plant because it unloads the coal-pulverizer unit. When the CCI project is located at a remote site, however, the coal-fines will need to be separated by screening and processed into briquettes using commercially available equipment. As noted, the criteria for selection of suitable feed-coal include evaluation of coal-fines production.

PYROLYSIS OIL AND GAS RECOVERY The recovery process for CTO and pyrolysis gas is presented in more detail in a separate paper (ref - 1), including a review of the compositions of various CTO fractions and several processing options. In summary, the volatilized coal-tar-oil (CTO) compounds are recovered from the pyrolysis reactor gas effluent by condensation. Several process configurations are feasible, however there are advantages to using a multi-stage condensing process that produces three or more distinct oil fractions, a process-water condensate, and a non-condensable fuel-gas fraction that is cleaned and used as fuel in the process. Due to the differences in composition between LRC and metallurgical coking-coals and different processing conditions (half-hour at 550°C versus 6-hours at 1000°C), the CTO composition from MTP is considerably different from coke-oven coal tar. For example, CTO has more low-boiling range material and less materials with boiling points above 500°C (930°F). The recovered CTO is suitable for catalytic hydrotreating to syncrude oil or additional fractionation into coal tar chemical intermediate products. To be economically viable, onsite syncrude oil production from CTO will require a capacity in excess of 7500-barrels per day corresponding to 10,000-t/d LRC feed and matching a 500-MW PC-power generating plant.

ANCILLARY PROCESS SECTIONS Several ancillary plant sections support the key process sections described above. These include water and gas recovery and cleanup, hydrogen production based on coal-gasification, cogeneration of electricity and steam, safety and emissions control systems.

A substantial amount of water is removed from the coal during the process. For example, a 10,000-t/d CCI plant supplying a 500-MW power plant and 7500-bbl/d syncrude oil based on "reference" LRC, will produce 3,300-t/d water or 137-tons/h (550-gpm). As a useful byproduct for boiler-feed water and cooling-tower water supply, the water is recovered and cleaned, removing all the contaminants, e.g., ammonia, mercury, organic compounds and sulfur compounds with conventional water treatment processes including ultra-filtration, gas stripping and adsorption.

The hydrogenation of CTO to syncrude oil requires hydrogen (approximately 1.7-t/h or 15-million scfd) generated from gasification of 250-t/d CCF followed with conventional conversion of syngas to hydrogen. Several gasification processes have been in commercial operation over the past twenty years including BG-Lurgi, ConocoPhillips, GE/Texaco, Shell and Siemens for production of syngas for hydrogen, ammonia, methanol, oxo-alcohols and acetic acid.

PROCESS DEMONSTRATION MATURITY The process is "developmental" in the sense that no commercial plants are in operation yet. However, the individual process elements and equipment have been operated in various industrial configurations. It is prudent to acknowledge technology risk associated with equipment scale-up and process optimization, and the process design therefore is based on commercially available modular equipment. The individual modular process sections are separate unit operations subject to independent engineering simulation, test operations and validation from equipment vendors. The technology risks therefore can be defined rigorously and managed with careful project design, vendor commitments and equipment testing as appropriate. This approach reduces and limits any project risks relating to equipment and process performance, product quality and yields, operating rate, and maintenance costs.

PROCESS MASS AND HEAT BALANCE

As an example of typical mass and heat balances for two different coals, the data in Table - 2 illustrate the variability of the yields of CCF and CTO based on the coal composition. Feed coal with less than 1.4w% sulfur and more than 30w% volatile matter generally perform well as feedstock because the CCF will be a clean-coal product and the oil yield will be economically sufficient.

As shown in Table - 2, approximately a third of the "volatile matter" remains in the CCF product, a third is recovered as CTO and the balance is pyrolysis gas used as fuel in the process. The CCF product retains all the ash and about 70% of the fuel value from the feed-coal, but only 50% of the weight. The recovered CTO has about 25% of the feed value and 12% of the weight. And the weight balance is water non-condensable low-btu-fuel gas, which is consumed in the process and comprises about 10w% and 4-% of the feed-coal thermal energy. A CCI project serving a 500-MW PC-power plant and using 10,000-t/d Buckskin mine coal will produce approximately 1150 t/d coal-tar-oil or syncrude oil, 5000 t/d coal-char-fuel and 3000-t/d water. The heat balance indicates the 17.7-GJ/t [16.8-mmBtu/t] feed-coal is converted into 23.6 GJ/t [22.4mmBtu/t] coal-char-fuel, and approximately 5-w% of the feed-coal is used in the process as fuel gas. The production of hydrogen for the syncrude hydrotreating process would use an additional 2 – 3w% of the feed-coal.

TABLE - 2
EXAMPLES OF MASS AND HEAT BALANCES

EXAMPLES *	1	2	
LRC feed	Subbituminous	Lignite	
Fuel Value, Btu/lb	8400	6400	
MillionBtu/t	16.8	12.8	
Volatile Matter, wt%	32	28	ASTM method
VM, WAF, wt%	49	52	Water & ash free basis
Moisture, wt%	30	41	Moisture excluding pyrolysis water
Ash, wt%	5	5	
MASS BALANCE	------------lb/t-feed-coal------		
Coal-char fuel	980 - 1030	800 - 860	
Coal-tar-oil	230 - 240	210 - 220	
Water	570 - 600	840 - 860	
Pyrolysis gas	190 - 210	180 – 190	
ENERGY BALANCE	--Million-Btu/ton feed-coal--		
Coal-char fuel product	11.3 – 11.9	7.3 – 8.5	
Coal-tar-oil	4.1 – 4.4	3.5 – 4.0	
Coal process fuel	0.8 – 1.0	0.8 – 1.1	
Total feed coal	16.8	12.8	

[*]These examples are based on generic coal data for illustration of the expected product yields Quantities from database in English units (with these conversion factors 0.454-lb/kg, 2000-lb/t, 1.055-kJ/Btu and 1.055-GJ/mmBtu) were used in this table.

COAL FEEDSTOCK SELECTION

The low-rank-coal (LRC) designation includes some bituminous most sub-bituminous coals, and lignite. The selection of feed coal for the CCI process depends partly on the required CCF quality, but more importantly also on the CTO yield that determines the economic viability as well as on several criteria relating to the physical parameters that define processability (Table – 3). To meet the CCF specifications, the process reduces the sulfur, nitrogen and mercury values according to the capability of each process step and the distribution of individual compounds in the coal. For example, 90% of the volatile mercury will be removed, but the reduction of sulfur depends on its distribution between volatile organic sulfur (50 – 75% removal), pyrite (35 – 75% removal) and sulfate (near zero removal). Thus, when an "emission compliance" clean-coal product is desired, the selection of the feed-coal must correspond to the final CCF specifications.

The physical processing characteristics of the feed-coal is important for processability and process design and is evaluated in terms of friability, coal fines production ratio, softening temperature and range, evolution rate and ratio of pyrolysis gas to condensable oil. Even the project capacity can be important in the selection of feed-coal, because cost effective conversion of CTO to syncrude or specialty chemicals will require a certain processing volume.

TABLE – 3
FEED COAL SELECTION CRITERIA

COAL PARAMETER	CRITERIA RANGE	COMMENTS
Volatile Matter, wt%	> 25	CTO yield correlation
Volatile Matter, WAF wt%	> 33	Determines CTO yield
CTO Yield, wt%	> 10	Economic determinant
Water content, wt%	15 – 55	Relates to process energy demand
Ash, wt%	< 15	Ash remains with the CCF product
Sulfur, wt%	< 1.5	Ref. S-distribution and CCF spec.
Softening temperature range, °C	-	*
Agglutinating value	6 – 11	*
Hardgrove Grindability index	30 – 90	*
Friability & strength at temp. -		*
Heating value, Btu/lb	-	*

[*] Subject to evaluation of specific coal

CLEAN COAL FUEL PRODUCT QUALITY

Compared with the feed-coal, the coal-char-fuel (CCF) has several important advantages for the PC-power generator, including lower emissions profile and mitigation cost, increased boiler efficiency and therefore reduced fuel requirement and CO2-emission (Table – 4). The sulfur is reduced with the pyrolysis gas, coal-tar-oil, and by separation of the pyrite-pyrrhotite (Table – 5).

TABLE – 4
COAL-CHAR FUEL ADVANTAGES

PARAMETER	ADVANTAGES & APPROXIMATE VALUES
Emissions	Lower SO_x by 40 – 70%, mercury by 85 – 95%
Emissions mitigation cost	Decreased operating costs
Heating value	Increased by 100-Btu per 1% water removal
Boiler efficiency	Increased by 0.16% per 1% water reduction
GHG emission, CO_2	Decreased by 0.16% per 1% water reduction

As indicated in Table - 5, the process can produce US-EPA 2010-CAIR-compliance CCF depending on the sulfur distribution between volatile organic compounds, sulfates and pyrite. The pyrolysis process removes volatile mercury and also nitrogen compounds in the same ratio as the sulfur reduction. Volatile mercury typically is present in LRC feed-coals at up to 200-wppb with atmospheric boiling point at 357°C, and it is therefore effectively removed from the coal matrix above the 550°C operating temperature resulting in 90% or better reduction.

TABLE – 5
COAL-CHAR-FUEL COMPOSITION – EXAMPLES

These examples are based on generic coal data for illustration from database in English units; (conversion factors 0.454-lb/kg, 2000-lb/t, 1.055-kJ/Btu and 1.055-GJ/mmBtu)

Example #	---------1--------		-------2---------	
Coal origin	Wyoming PRB		New Mexico	
Coal type	Subbituminous		Lignite	
	Feed	CCF	Feed	CCF
Heating value, Btu/lb	8400	11200	6400	8700
Data ranges	-----------------------------wt%---------------------------			
Volatile Matter	27 – 33	14 – 20	25 – 28	12 – 18
Water	28 – 32	nil – 2	35 – 45	nil – 2
Ash	5 – 7	7 – 12	6 – 8	11 – 14
Sulfur type:				
Pyrite	0.12	0.06	0.12	0.06
Sulfate *	0.10	0.15	0.11	0.22
Volatile organic	0.48	0.25	0.42	0.21
Total sulfur	0.70	0.55	0.65	0.53
	-----------------S-lb/mmBtu--------------------			
Sulfur, lb/mmBtu	0.71	0.31	0.84	0.36
SO_2, lb/mmBtu **	1.43	0.62	1.68	0.72

[*] Sulfate reports to slag or fly-ash and therefore is not included in SO_2-lb/mmBtu
[**] The US.EPA 2005 CAIR rule for 2010 SO2-compliance coal is max. 0.62-lb/mmBtu

As shown in the examples in Table - 5, removal of water, pyrolysis gasses and oils from LRC results in 40% or more increased heating value. This translates directly to an increased boiler efficiency of 1% per 5 – 6w% water removed, and an important commensurate reduction in fuel, combustion air and GHG emissions. Test burns using dried coal and pyrolysis coal-char-fuel in utility boilers have confirmed satisfactory ignition, flame temperature increases of about 100°C, and good flame stability and burnout.

POWER GENERATION PROJECT ADVANTAGES

For a PC-generating power plant located near the CCI coal conversion project, there are several prospective advantages (Table – 6 and Exhibit - 1) in addition to the ones noted above.

TABLE – 6
COAL-CHAR-FUEL AND CCI ADVANTAGES FOR PC-POWER PLANT

PARAMETER	IMPROVEMENT AND CAUSE
Emissions	Reductions due to decrease in SO_2, NO_x, Hg
Heat-rate improvement	Decrease due to fuel heating value upgrade
GHG improvement	Reduction in CO2 due to heat-rate decrease
Captive power	Decreases due to less combustion air flow, feed-coal weight, etc.
Water supply	Increase resulting from coal moisture recapture
Stack plume	Decreased stack gas plume due to reduced water content
Feed-coal flexibility	Expanded flexibility to use lower rank coals as fuel
Other advantages	Prospective synergism from CCI and PC-generation
[IGCC efficiency	Efficiency gain provides basis for using LRC in IGCC projects]

Due to the significant improvement in thermal efficiency offered by processing LRC for IGCC and CTL projects, the CCI process may become an important enabling technology for LRC projects that otherwise might not be economically viable. Prospectively this may lead to LRC becoming competitive with high-rank coals for the next generation of CTL projects.

COAL-TAR PRODUCTS AND CONVERSION TO SYNCRUDE

The coal tar-oil (CTO) composition and conversion to syncrude or coal-tar chemicals is a complex and important feature of the CCI process, and is presented in more detail in a separate paper (ref - 1). In summary, the properties of CTO produced with mild-temperature pyrolysis (Table – 7) are different from those of coal-tars derived from metallurgical coke ovens and town-gas plants. This is due to the different composition of coking coals versus LRC, the homogeneous heating and relatively short residence time preventing excessive deterioration of the lower boiling liquids, and the relatively mild pyrolysis temperature limiting the volatilization of the 650°C (1200°F) boiling range compounds. The low partial pressures of the individual CTO compounds during pyrolysis also assist in preventing polymerization and other reactions. The CTO therefore has properties more suited for conversion to chemical intermediates by distillation and extraction, and to syncrude oil by catalytic hydrotreating. As described in a separate paper, CTO can be hydrotreated using conventional refinery processes, yielding a 20 – 25% lower density syncrude oil product with correspondingly higher volume and a density of about 0.90 g/ml (0.26-API Gravity).

Table 7 Comparison of Properties of Coal Tar Oil and SynCrude Oil

Numeric data ranges:	**Coal-Tar-Oil Range** Example for illustration	**SynCrude Oil Range** Example for illustration
Product:	Coal-Tar-Oil	Syncrude Oil
Process	Pyrolysis Oil	Hydrotreated
Specific Gravity, g/cc	1.04	0.85 – 0.95
Pour point, °C	22 – 35	-10 – -12
Distillation range, °C	150 – 700	to be determined
Coal particles, wt%	0.2 – 0.5	0.2 – 0.5
Elemental analysis:	wt%:	
C	78 – 80	
H	8.0 – 9.2	
S	0.9 – 1.5	
O	6.5 – 9.5	
N	0.3 – 0.9	
Compound group:		
Cresylic acid fraction	8 – 10	
Amine fraction	2 – 3	
Hydrocarbons C6 – C14	5 – 8	diesel fraction
Hydrocarbons C15 – C32	30 – 40	wax fraction
Hydrocarbons C33 – C44	35 – 45	pitch fraction

SYNCRUDE OILFIELD DEVELOPMENT PERSPECTIVE

In view of the current cost of crude oil and the desire to increase domestic supply, it is tempting to evaluate what contribution the CCI process could make in this regard. First, LRC feedstock for the process is abundantly available in the U.S. for the foreseeable future well in excess of several hundred years, and the present LRC mining capacity exceeds 700-million tons per year serving more than 200 major coal power generating plants. Second, the economic viability of CCI processing of LRC appears to be attractive based on current coal and oil prices, and a single CCI "reference project" costing about $500-million and serving a 500-MW PC-generating plant would co-produce about 7500-bbl/day syncrude after hydrotreating of CTO on-site.

Third, an oilfield producing 100,000-bbl/day with more than 20 years of recoverable oil in place (ROIP) is considered to be a "major oil resource" using the terms used in the oilfield exploration-&-production (E&P) industry. To rank as the equivalent of a "major oilfield discovery", just fourteen CCI nominal 10,000-t/d projects would need to be in operation serving 7000-MW of PC-power or IGCC generating capacity. In terms of project capital intensity, the CCI reference project is similar to the power generating industry at $1400 per kW capacity or $9.8-billion for 7000-MW. Perhaps more importantly, the CCI project also has a favorable capital cost ratio compared with the oil E&P industry at $15 or more per barrel ROIP (ref – 4) or $10.5-billion for a 20-year 100,000-barrel/day oilfield, (Table – 8).

TABLE – 8
ENERGY PROJECTS CAPITAL COMPARISON

PROJECT TYPE	CAPACITY	CAPITAL COST PARAMETER (U.S. basis 2006 cost estimates)	TOTAL ($$10^6$)	CAPITAL RATIO
PC-Power generation	500-MW	$1400/kW, 20-yr life	700.0	1.00
Oil field development	7500-bbl/d	$15/bbl-ROIP **, 20-yr life	787.5	1.13
CCI Syncrude project	7500-bbl/d 5000-t/d clean-coal	$11/bbl-ROIP **, 20-yr life	577.5	0.83

[*] Capital cost is provided as generic examples depending on project location, capacity, etc..
[**] Recoverable oil in place [ROIP] based on any given oil recovery technology

An evaluation of the CCI project as an "equivalent virtual oilfield" would include as assessment of the comparative project risk factors (Table – 9) and net-present-value (Table – 10). The CCI project risks compare favorably with oilfield E&P projects for several reasons as indicated. This suggests that such LRC based CTL projects are feasible and could have wider application. For example, using as a basis for CCI project development just one third of currently mined LRC in the U.S. or 250-million-t/y coal serving seventy 500-MW-capacity power plants, the resulting new domestic syncrude oil resource would produce at the rate of 525,000-bbl/d. The additional benefits to the power generating sector already have been mentioned.

The cost and risk of oilfield exploration & production (E&P) projects are known to vary greatly according to the particular geology, oil quality, amount of "recoverable oil-in-place" (ROIP), cost of production and transportation, climatic conditions, offshore concerns including storms, country related political and currency stability, etc. The CCI project, however, when viewed as an oilfield project located near a domestic power-generation plant presents a much reduced risk profile. Thus, the syncrude oil quality, ROIP and production rate are all definable in advance based on specific plant design and coal supply contracts. And other important oilfield project risks appear to be absent altogether including risks associated with international shipping, offshore platform operation, storms, and country related political, contractual and currency instability. It therefore seems that CCI projects could have important operational and financial advantages as means for oilfield resource development.

TABLE – 9
CCI AND OILFIELD PROJECT COMPARATIVE RISK FACTORS

EVALUATION OF PROJECT RISK FACTOR FOR OILFIELD E&P PROJECTS

PROJECT: RISK FACTOR	OFF-SHORE OILFIELD RISK ASSESSMENT	CCI PROJECT RISK ASSESSMENT
Crude oil market pricing	Major uncertainty	Major uncertainty
ROIP* uncertainty	Medium to high	Low or none
Geology uncertainty	Medium	Low or none
Recovered oil quality	Medium	Low or none
Exploration cost	Uncertain	Low and defined
Project capital cost	Uncertain	Defined
Project technology risk	Defined or definable	Definable **
Project operating risk	Low	Low or none**
Production rate	Definable but uncertain	Defined
Production cost	Definable but uncertain	Defined
Transportation cost	Uncertain	Defined
Inclement weather risk	Uncertain	Low or none
Political stability	Definable risk	Low or none
Contractual stability risk	Definable risk	Low or none
Currency exchange rate risk	Uncertain	None
Coal supply cost	Minor or none	Minor or none

[*] Recoverable oil in place [ROIP] based on any given oil recovery technology
[**] CCI technology risk has some uncertainty until several plants are in operation

TABLE – 10
OILFIELD VALUATION COMPARISON

COMPARISON OF OILFIELD PROJECT NET PRESENT VALUES

PROJECT:	OFF-SHORE OILFIELD	CCI PROJECT
Capacity, (4 x 7500-bbl/d)	30,000-bbl/d	30,000-bbl/d
Number of operating wells or units	10	4
Risk evaluation profile	Complex	Definable
Operating rate	350-d/yr at 100%	350-d/yr at 100%
Capital cost (ref. 350-d/yr x 20-yr)*	$15/bbl-ROIP	$11/bbl
Total project capitalization	$3150-million	$2310-million
First year EBITDA, $50/bbl	$525-million	$525-million
EBITDA NPV **	$4277-million	$4277-million
EBITDA NPV less investment	$1127-million	$1967-million

[*] Capital cost is provided as generic examples depending on project location, capacity, etc.
[**] EBITDA, NPV (10% discount rate, 20-years) based on constant price of oil.

PROSPECTIVE PROJECT ECONOMICS

Efforts to take technologies for clean coal enrichment and coal-to-liquids past the demonstration stage and into commercialization have been impeded in the past by low market prices for oil. However, using the current and forecasted prices for candidate coals and competing refinery feedstocks, the current economics look to be favorable. Early forecasts of the economic results indicate probable rates of return on capital in the range of 25-30%. To provide a valid project economic evaluation it would be necessary to define the feed-coal, product yields for CCF and CTO, project capacity, capital cost and costs associated with the project site and project financing. However, a generic project example may help illustrate the general sense of the economic expectations (Table-11). Economic feasibility evaluations of a CCI project based on a conceptual reference plant have been prepared. The project economic example is based on a reference design plant capacity of 10,000 t/d feed-coal using Wyoming PRB Wyodak coal, with four parallel 2500-t/d pyrolysis reactor modules operating in parallel. The project can supply CCF to a 500-MW PC-power station and co-produce 8000-bbl/d syncrude, providing sufficient CTO to allow on-site hydrotreating to be economically viable. The operating rate, capital assumptions and other economic data are summarized in Table - 11, indicating an attractive result.

TABLE – 11
CCI PROJECT ECONOMIC EXAMPLE

PROJECT ECONOMIC EVALUATION PREMISES AND EXAMPLE OF COSTS

Capacity, feed-coal, t/d		10,000	16.8-mmBtu/t	$20/t	$1.19/mmBtu
Clean-coal product, CCF, t/d		5000	22.4-mmBtu/t	$38/t	$1.70/mmBtu **
Syncrude oil, t/d, 26-API.Gr		1310	8200-bbl/d	$50/bbl	
Operating rate,		350-d/y			
Operating cost,		$million/y			
	Feed-coal	70.0			
	Other operating cost	19.4			
	Total O&M cost	89.4			
Sales:	Syncrude oil	143.5			
	Clean-coal CCF	66.5			
	Total sales	210.0			
EBITDA		120.6			
Total capital cost		$500-million, parametric			
Ratio, EBITDA/Total capital cost		0.241	without fuel tax-credits, financial charges and taxes		

[*] Capital cost is provided as generic examples depending on project location, capacity, etc.
[**] Assumed coal price is $20/t and CCF price is coal cost times 1.43 x coal cost per Btu/lb

CONCLUSIONS

The CCI pyrolysis process is technically feasible based on past demonstration plant operations and process design improvements integrating commercially available equipment used in similar industrial applications. Although design scale-up to commercial capacity does not present any unusual or major engineering challenges, however the technology remains developmental until the first commercial project is in operation. Because the U.S. and worldwide LRC resources are abundant, the conversion of LRC to low-emissions "2010 CAIR-compliance clean coal" is attractive, in particular when the simultaneous co-production of syncrude oil can provide a major domestic crude oil resource at comparatively low risk. Based on current LRC production and use for power generating, the CCI process can make a meaningful contribution to the domestic supply of syncrude oil of the order of 8000-bbl/d for each 500-MW power plant. For the Western LRC producers the process provides important opportunities due to the significant increase in energy efficiency resulting from the clean-coal product when used in PC-generating plants and new IGCC projects that otherwise may turn to Eastern coal supplies and new emissions mitigation technology. In addition to being a prospective enabling technology for IGCC projects based on LRC, the CCI process also expands the market for LRC as an additional 15% of the power plant feed-coal is converted into syncrude. The CCI process appears to be sufficiently attractive to provide strong incentives for early commercial development based on the promising process economics, present energy prices and wide interest in developing domestic resources of syncrude oil,

REFERENCES

1 E. R. Skov, et al, "Coal-Tar Chemicals and Syncrude Oil Production from Low-Rank Coals Using Mild-Temperature Pyrolysis", AIChE Spring National Meeting, Houston, Texas, April 22 – 26, 2007.

2 John B. Goodman, et al, "Laboratory Carbonization Assay of Low-Rank Coals at Low, Medium, and High Temperatures, U.S. Bureau of Mines Report of Investigations 5383, January 1958.

3 Hydrotreating, Phillips Petroleum Lab., using HTAS TK-555 at 650°F, 900-psia.

4 Oilfield acquisitions range is $8 – 34/bbl-Recoverable-Oil-In-Place: Oil&Gas Journal and Wall Street Journal, e.g. WSJ Apr-11 and 19, 2006.

5 J.F. Jones, et al, "Char Oil Energy Development, Volume I", September 1975, U.S. Department of Commerce Report FE-1212-T-9 (NTIS), FMC Corporation, Princeton, NJ.

6 E. R. Skov, et al, "Low Rank Coal Upgrading with Syncrude Oil Production Using "SynCrude-SynCoal™" ….", 23rd Annual International Pittsburgh Coal Conference, Pittsburgh, PA, September 25 – 28, 2006.

END..//

FIGURES ATTACHED

Process flow schematics

1 Process Overview Block Diagram
2 PFD-100, Coal Drying & Pyrolysis
3 PFD-200, CTO Recovery
4 PFD-300, Hydrotreater

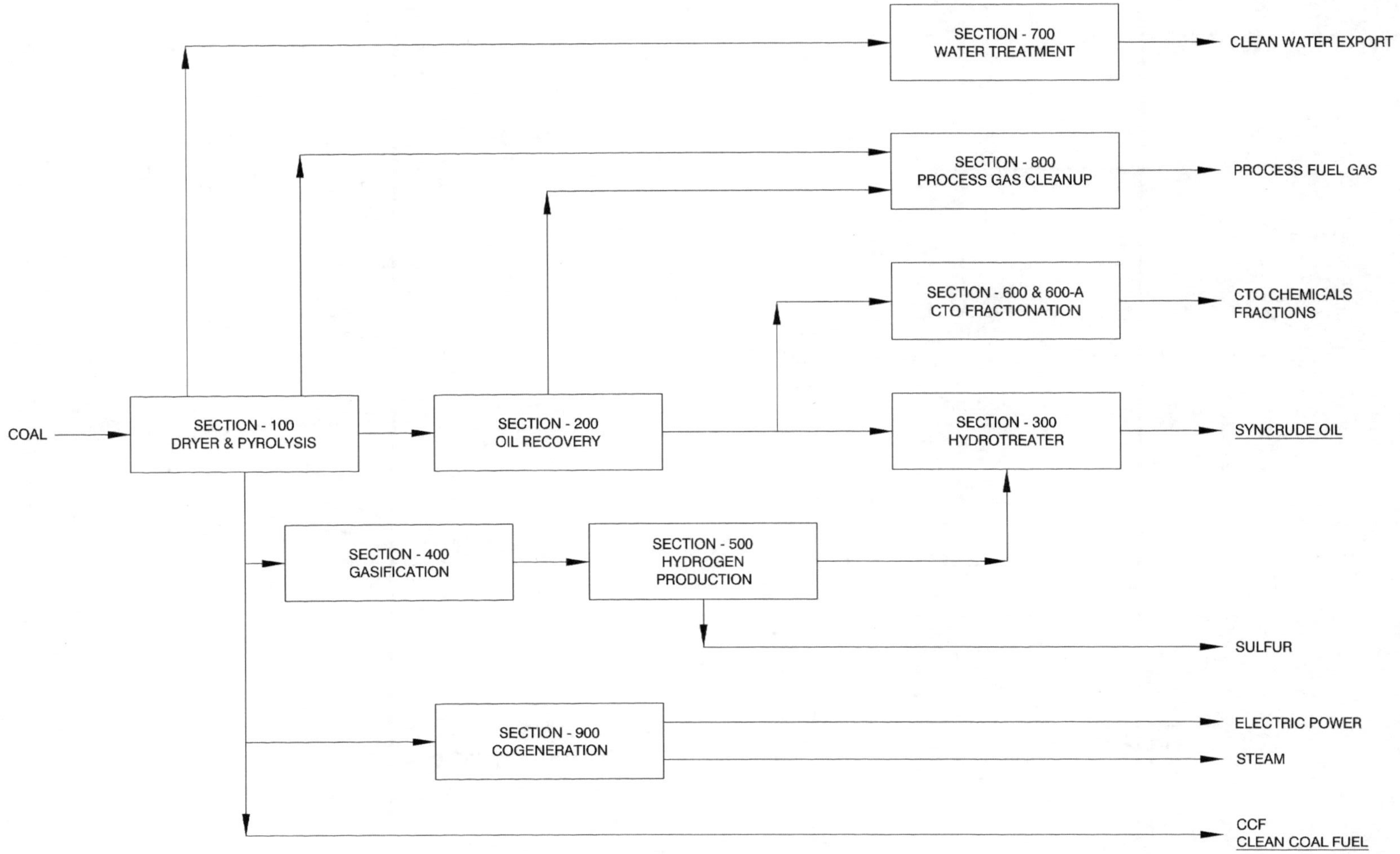

ConvertCoal, Inc.

SYNCRUDE-SYNCOAL
PROCESS SUMMARY

Revision Date - 03/09/2007

T:\2000's\2026\2026.001\008 - 03/09/2007

T:\2000's\2026\2026.001\001 - 03/12/2007

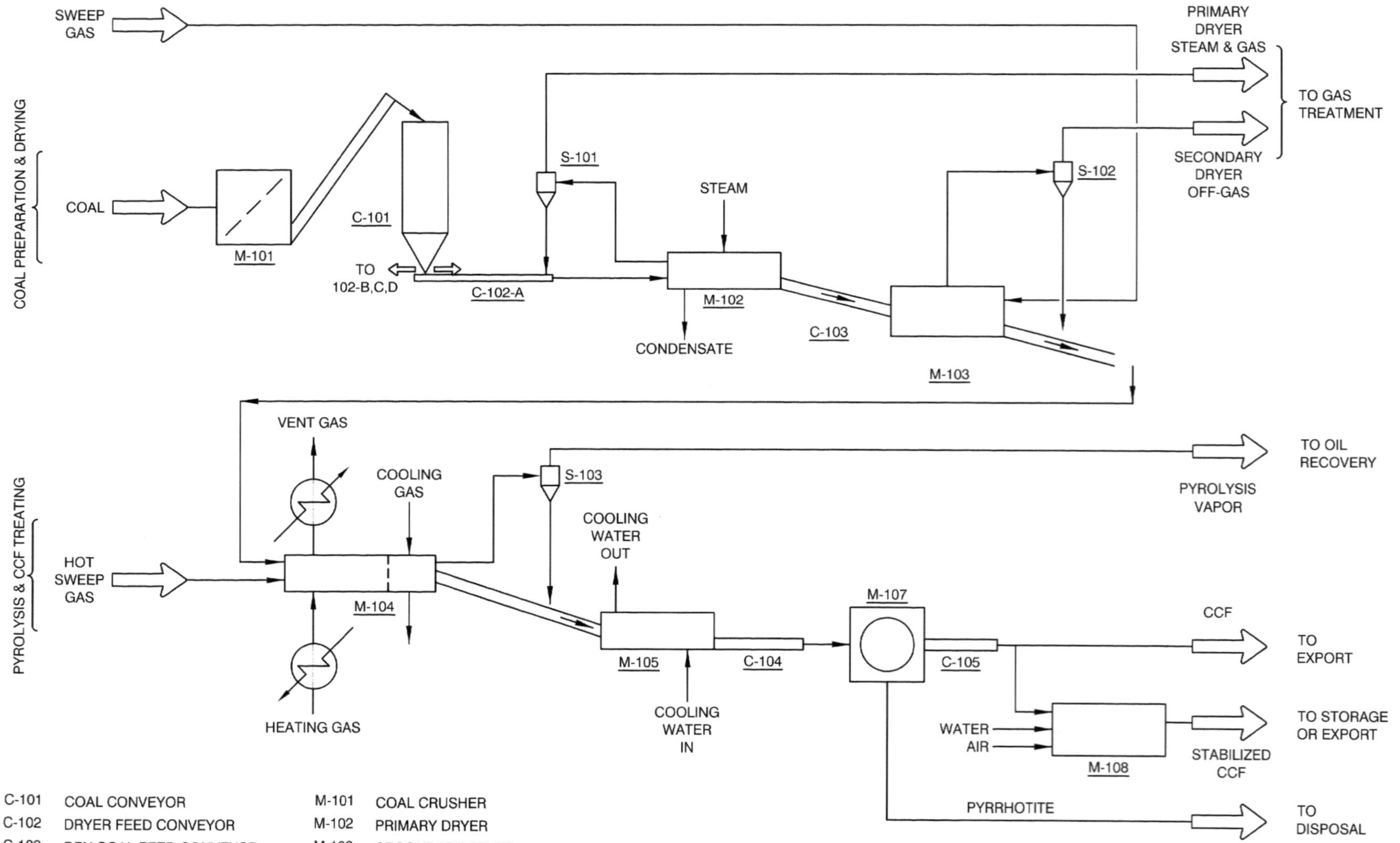

C-101	COAL CONVEYOR	M-101	COAL CRUSHER
C-102	DRYER FEED CONVEYOR	M-102	PRIMARY DRYER
C-103	DRY COAL FEED CONVEYOR	M-103	SECONDARY DRYER
C-104	COOL CCF CONVEYOR	M-104	PYROLYZER
		M-105	CCF COOLER
S-101	COAL-FINES SEPARATOR NO. 1	M-106	CCF CRUSHER
S-102	COAL-FINES SEPARATOR NO. 2	M-107	PYRRHOTITE SEPARATOR
S-103	COAL-FINES SEPARATOR NO. 3	M-108	CCF STABILIZER
Y-101	COAL SILO		

ConvertCoal, Inc.

PROCESS FLOW DIAGRAM 100

COAL DRYING & PYROLYSIS

Revision Date - 03/12/2007

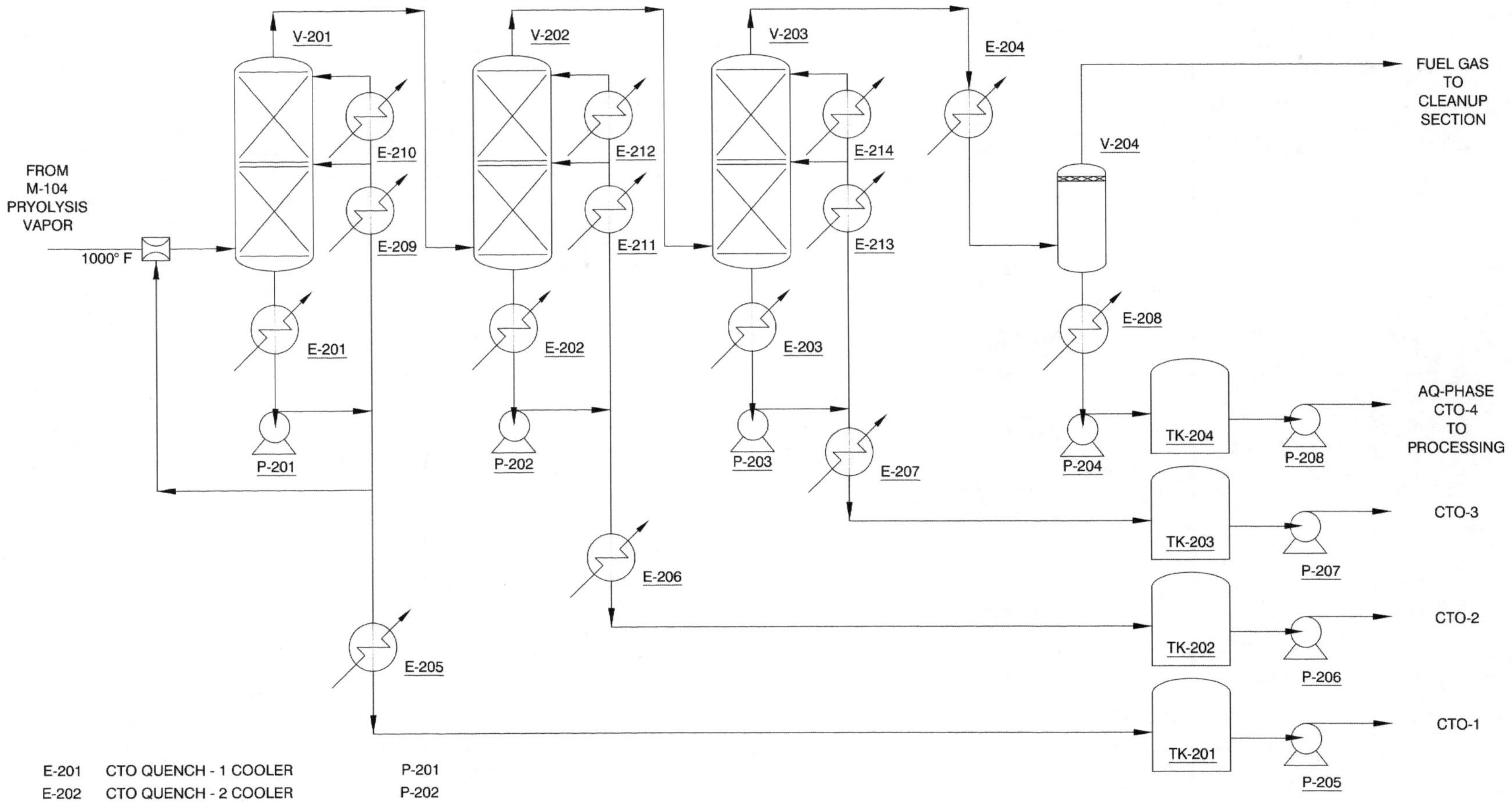

E-201	CTO QUENCH - 1 COOLER	P-201	
E-202	CTO QUENCH - 2 COOLER	P-202	
E-203	CTO QUENCH - 3 COOLER	P-203	
E-204	WATER CONDENSER	P-204	
E-205	CTO - 1 PRODUCT COOLER	P-205	
E-206	CTO - 2 PRODUCT COOLER	P-206	
E-207	CTO - 3 PRODUCT COOLER	P-207	
E-208	CTO - 4 PRODUCT COOLER	P-208	
E-209	Q-1 TRIM COOLER - 1		
E-210	Q-1 TRIM COOLER - 2	V-201	VAPOR QUENCH - 1
E-211	Q-2 TRIM COOLER - 1	V-202	VAPOR QUENCH - 2
E-212	Q-2 TRIM COOLER - 2	V-203	VAPOR QUENCH - 3
E-213	Q-3 TRIM COOLER - 1	V-204	VAPOR QUENCH - 4
E-214	Q-3 TRIM COOLER - 2		

ConvertCoal, Inc.

PROCESS FLOW DIAGRAM 200

CTO RECOVERY

Revision Date - 03/09/2007

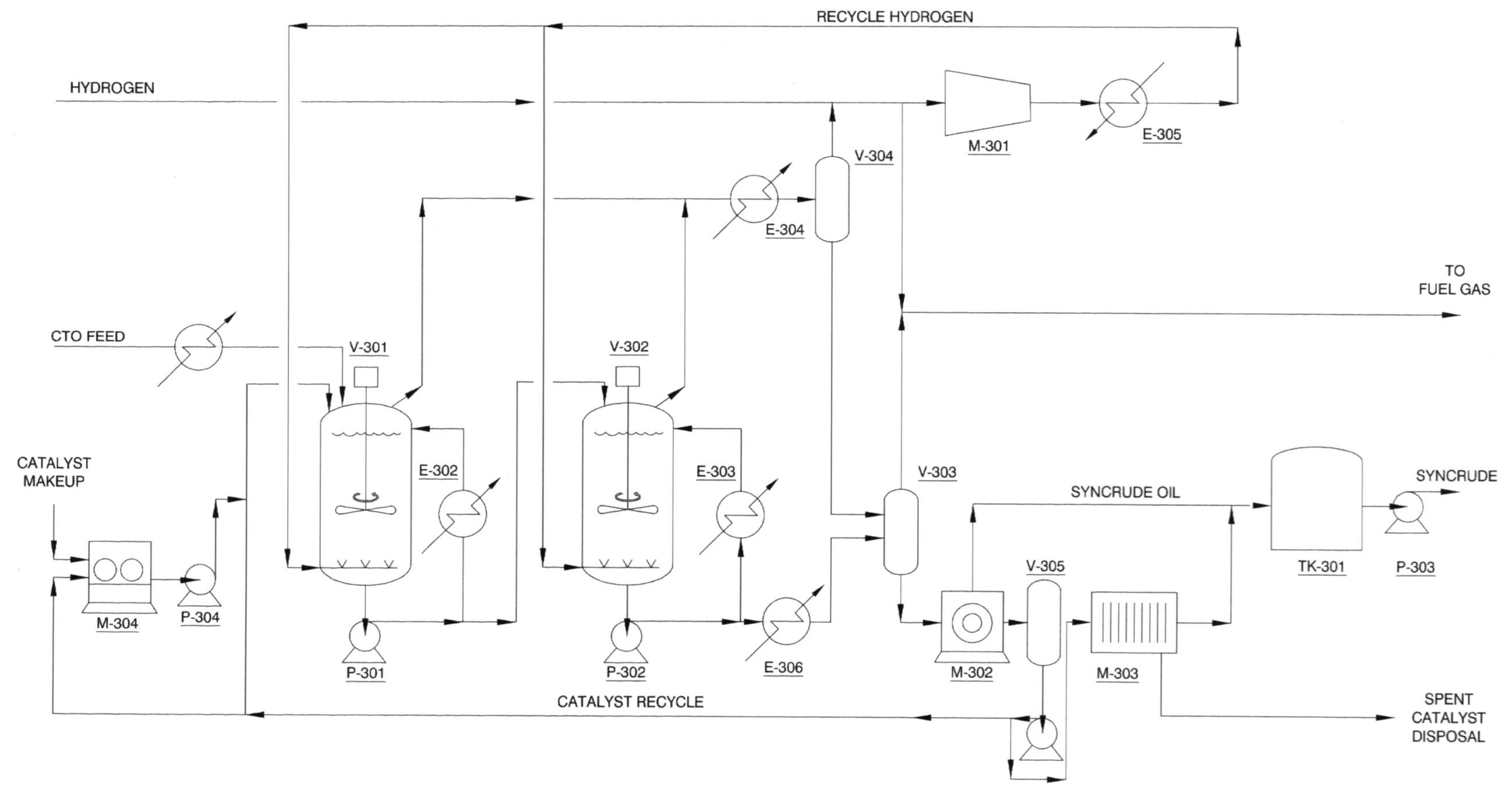

V-301	HOT-SLURRY REACTOR - 1	E-301	HOT-FEED HEATER
V-302	HOT-SLURRY REACTOR - 2	E-302	HOT REACTOR-1 COOLER
V-303	DEGASSING VESSEL	E-303	HOT REACTOR-2 COOLER
V-304	COMPRESSOR KNOCK OUT DRUM	E-304	RECYCLE HYDROGEN COOLER
V-305	CENTRIFUGE RECEIVER	E-305	RECYCLE HYDROGEN HEATER
M-301	HYDROGEN RECYCLE COMPRESSOR	E-306	SYNCRUDE OIL COOLER
M-302	CATALYST RECOVERY CENTRIFUGE		
M-303	CATALYST FILTER	P-301	HOT-REACTOR-1 RECYCLE PUMP
M-304	MAKE-UP CATALYST MIXER	P-302	HOT-REACTOR-2 RECYCLE PUMP
		P-303	SYNCRUDE PUMP
TK-301	SYNCRUDE OIL TANK	P-304	MAKE-UP CATALYST PUMP

ConvertCoal, Inc.

PROCESS FLOW DIAGRAM 300

HYDROTREATING OF CTO SLURRY PROCESS

Revision Date - 03/09/2007

T:\2000's\2026\2026.001\003 - 03/09/2007

Novel Warm Synthesis Gas Cleanup for Igcc Systems

Brian Turk[1], **Raghubir P. Gupta**[1], and Jerry Schlather[2]. (1) Center for Energy Technology, Research Triangle Institute, 3040 Cornwallis Road, Johnson Building, Research Triangle Park, NC 27709, (2) Eastman Chemical Company, Kingsport, TN 37662

For last 15 years, RTI's Center for Energy Technology (CET) has been working with the Department of Energy (DOE) to develop technologies to support DOE efforts in coal gasification for production of electric power and value-added products. A unique technology platform has been developed to remove sulfur and other contaminants from syngas derived from gasification of coal and other carbonaceous feedstocks. This development work not only includes the development of unique materials for removal of various contaminants in the syngas, but also integration of these removal processes into overall integrated gasification combined cycle (IGCC) system. Compared to existing technologies, these new technologies will reduce capital and operating costs of syngas clean-up significantly. With use of these new processes, the overall thermal efficiency of power generation from coal will be improved, which will also result in lower carbon emissions. Also, coal will be able to replace natural gas as a feedstock for energy-intensive production of commodity chemicals, like ethylene, hydrogen, etc. Overall, this technology package will improve the competitive position of coal gasification versus conventional pulverized coal combustion for power generation. This paper will present the status of technology development efforts (primarily the results of pilot-plant testing with real coal derived syngas at Eastman Chemical) and technical and economic advantages of this technology compared to Rectisol and Selexol processes.

Web Page: www.rti.org/energyreserach

Preliminary Study of the Gasification Process of Mexican Fuels at Atmospheric Pressure

Jorge A. Altamirano-Bedolla, Alejandro Mani-González, Emilio Manzanares-Papayanopoulos*, Manuel Fernández-Montiel, and César A. Romo-Millares

Gerencia de Procesos Térmicos, Instituto de Investigaciones Eléctricas, Reforma No. 113, Col. Palmira, CP62490, Cuernavaca, Morelos, México. ****Speaker, e-mail: emp@iie.org.mx***

ABSTRACT

The Instituto de Investigaciones Electricas (IIE) has been working in the past years on economical and technical evaluation studies of the Gasification Technology as an alternative for power generation. Despite being considered as an expensive technology compared to other technologies for power generation, the gasification technology as Integrated Gasification Combined Cycle (IGCC) become attractive in many ways since it offers a clean path to power generation, also a flexibility on the feedstock, in which let open the fuel diversification to be used in power plants.

The main energy company in Mexico, Comision Federal de Electricidad (CFE), is interested in the potential that the gasification technology offers for power generation in the following 5 to 10 years. Also, the main oil company, Petroleos Mexicanos (PEMEX), is interested in the gasification technology for power generation, as well to obtain from the syngas feedstocks, i.e. hydrogen, for its refinery processes.

In that sense, IIE have planned an ambitious research and technological development project to gather the knowledge of the technology and to gain expertise, in particular on the main topics related to (1) the thermodynamics and kinetics of the chemical reactions inside of the gasification reactor, (2) hot clean-up syngas systems, (3) solid and liquid residual feedstock systems, and (4) modelling processes using Computer Fluid Dynamic (CFD) tools.

The project has been structured in various stages aiming to design a research pilot plant gasifier. As partial result, a study of the behaviour of the syngas production in function of the composition of the fuel, temperature and pressure using a computational commercial program with a general model of the gasification process have been presented previously.

At this stage, a preliminary experimental study at work-bench scale of syngas production in function of the composition of the fuel and temperature at constant atmospheric pressure has been carried out using an electric furnace in a batch process. The fuels used in the experimental work have been a bituminous type C coal (ASTM, 2005) from MICARE mine at the north of the country with relative high ash concentration, and refinery liquid residual with relative high sulphur and metals concentration. The results obtained allowed us to design the experimental prototype of a continuous gasification system at work-bench scale.

As parallel stage, a CFD models for the gasification thermodynamics and kinetics of the chemical reactions of Mexican feedstocks used have been developing. The experimental results will validate the computational models.

I. INTRODUCTION

In México, about 76% of the electric power generation is based on fossil fuels. Furthermore, the electric power generation for the next 10 years is expected to continue with the use of such kind of fuels (SENER, 2006).

The expansion around the world and also in Mexico of the electric power generation in the last ten years has been based on natural gas combined cycle power plants. However, the volatility of the natural gas price and the worldwide environment concerns of clean power generation point out a clear path to explore new technologies for electric power generation, such as the gasification technology.

The Instituto de Investigaciones Electricas (IIE), technological research arm of the Energy sector in Mexico, has followed on past years the status of the gasification technology as Integrated Gasification Combined Cycle (IGCC) for power generation (Fernández-Montiel and Alanis, 1993 and 1994, Fernández-Montiel, 1994; Fernández-Montiel, *et.al.*, 2003). This interest of the gasification technology has been because it offers flexibility of the fuel used and with a required pollution reduction systems allows clean power generation (Higman and Der Burgt, 2003; Rezaiyan and Cheremisinoff, 2005).

Actually, the IIE has turn to gather the knowledge and to gain expertise in particular of the black box of the gasification technology as the gasifier reactor, which is considered as the heart of the technology. Nonetheless, the peripherical systems as the feedstock systems, syngas clean system, etc., have been considered for research studies.

As first stage of the IIE gasification technology main project, a study of the analysis of the syngas production as a function of fuel compositions, temperatures and pressures by simulation carried out by commercial computational packages and the comparison of the energy produced as function of the fuel composition at the same operational parameters have been presented previously (Manzanares-Papayanopoulos *et. al.*, 2006).

The work presented here is a second stage of an ambitious research and technological development project on the gasification technology for the main energy companies in

Mexico. One of the goals of the work is the feasibility study of the use of a feedstocks produced in the country. Nonetheless, some imported feedstocks as Petcoke and coal is also explored.

The aim of this study is to present preliminary experimental results of a batch gasification process of coal and refinery liquid residuals at work-bench scale. The results obtained have allowed to design a continuous gasifier including a solid and liquid feedstocks systems, temperature control system, detection syngas system, at work-bench scale. It is important to mention that this continuous gasification system is actually under construction.

II. EXPERIMENTAL

The main goal of the gasification experiments carried out in an electrical furnace was to observe and to analyse the three main stages identified in the gasification process: drying, devolatilization and gasification of Mexican feedstocks.

In order to achieve this goal, an experimental gasification system at work-bench scale was design and constructed. The gasification system was composed by the following configurations:

- Heat system, with an electric furnace capable to reach to up 1000 ºC (1832 ºF), variac voltage control, and a temperature sensor system.
- Monitoring syngas system, with a gas analyzer based of electrochemical cells for O_2, CO and NO_x, and a temperature gas sensor system.
- Data recording system, with a digital video record system, television, lamps and chronometers.
- Confine system, an acrylic box with a glass top to keep a defined area to perform the experiments.

(a) **(b)** **(c)**

Figure 1. Experimental gasification system, (a) electrical furnace, (b) view of the heat, data recording and the confine systems, and (c) same as (b) with the monitoring syngas system.

Figure 1 shows the experimental gasification equipment used during the preliminary gasification experiments of Mexican feedstocks.

III. RESULTS AND DISCUSSION

The experimental procedure was carried out in a batch process. A brief description of the procedure is described next. The fuel sample was placed inside of a vessel which was previously dried and weighed. Therefore, the sample was also weighed and placed in the electric furnace as fixed bed. Heat was applied to the sample from ambient temperature to up 1000 °C.

The CO concentration was monitoring by a gas analyzer based of electrochemical cells over 10 cm from the syngas output, also the temperature of the syngas was measured during the experimental runs. The whole process was recorded by a digital video camera, and projected on a colour television in which the gasification phenomenon was analyzed. The confine system was open therefore air with oxygen content about 21%v without any supply control was used as oxidant agent ,and the pressure was atmospheric.

The fuel tested were bituminous coal type "C" (ASTM, 2005) from the Mexican Micare mine at the north of the country and a refinery liquid residual from Tula refinery. The proximate analysis and weights are listed in Table 1. Also, the conversion from the initial sample and the remaining sample were calculated and listed in Table 1.

Table 1. Proximate analysis and weights of the fuel tested.

Property	Tula Refinery Liquid Residual		Micare Coal Sample 1		Micare Coal Sample 2	
	wt %	grs.	wt %	grs.	wt %	grs.
Ash	0.13	0.0044	38.20	1.4041	38.20	1.1961
H_2O	0.10	0.0034	1.25	0.0459	1.25	0.0391
Volatiles	--		26.08	0.9586	26.08	0.8166
C	--		34.47	1.2669	34.47	1.0793
Volatiles + C	99.77	3.3657	60.55	2.2255	60.55	1.8959
Conversion	**98.54 %**		**70.86 %**		**88.94 %**	

Figure 2 shows the behavior of the experimental test of Micare coal sample 1. The inset (a) shows the coal sample, (b) and (c) shows the devolatilization stage, and insets (d) and (e) shows the gasification stage of the experiment, and finally (f) shows the remaining ashes. Similarly, Figure 3 shows the behavior of the experimental text of Tula refinery liquid residual.

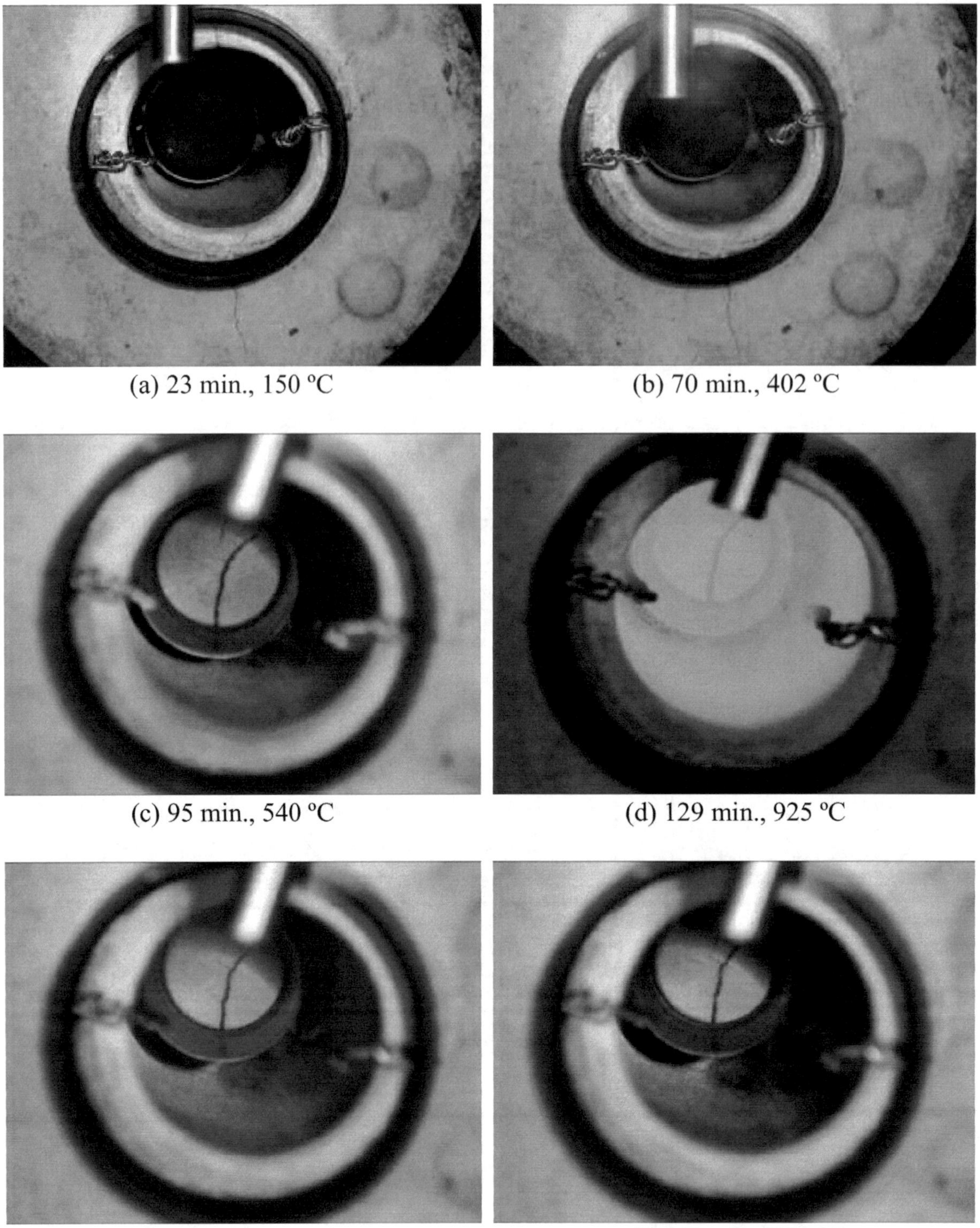

(a) 23 min., 150 °C

(b) 70 min., 402 °C

(c) 95 min., 540 °C

(d) 129 min., 925 °C

(e) 144 min., 835 °C

(f) 163 min., 478 °C

Figure 2. Behavior of the experimental test of Micare coal sample 1.

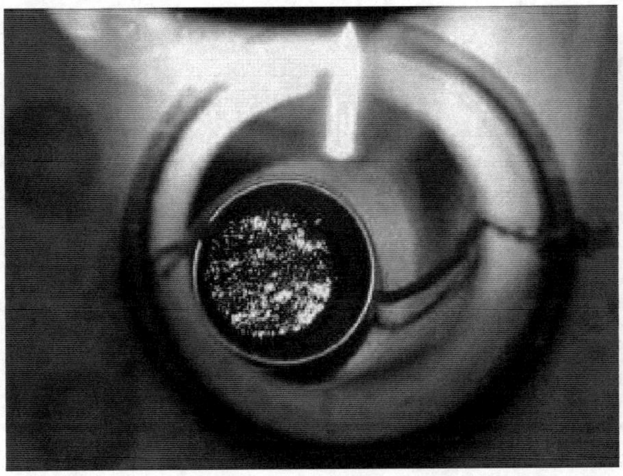

(a) 34 min, 215 °C

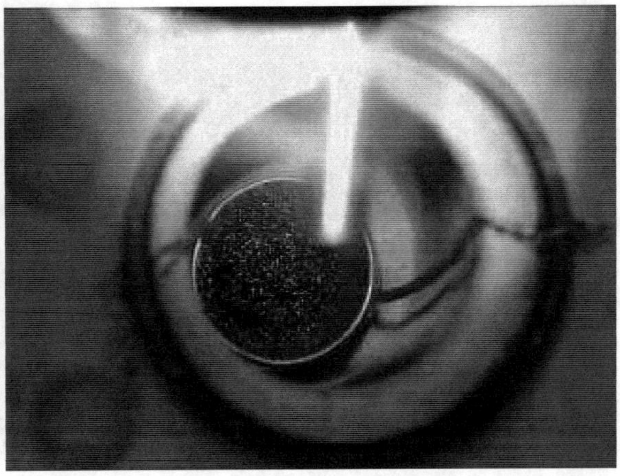

(b) 56 min, 334 °C

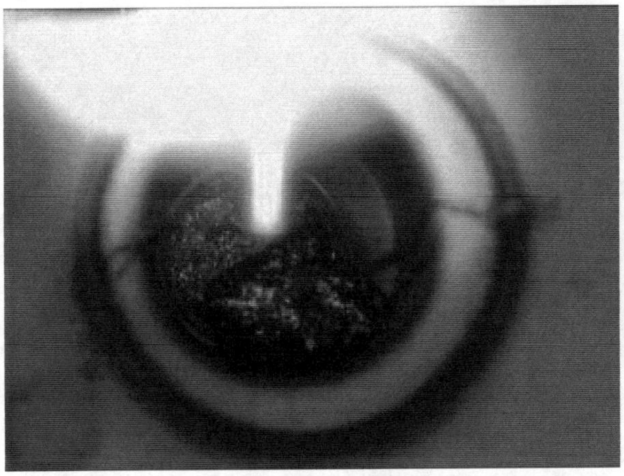

(c) 85 min, 380 °C

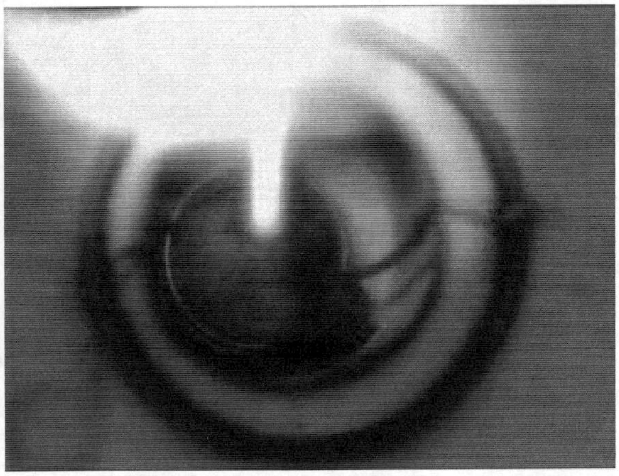

(d) 116 min, 485 °C

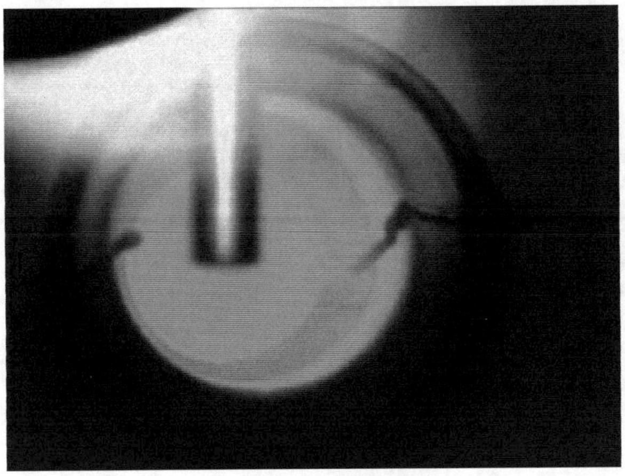

(e) 192 min, 988 °C

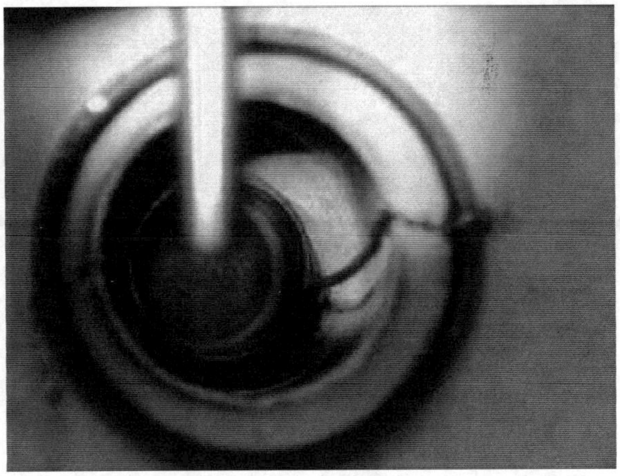

(f) 231 min, 447 °C

Figure 3. Behavior of the experimental test of Tula refinery liquid residual.

The CO behavior was monitored during the test for all the fuels tested. The behavior was similar for the experiments performed for the coal and refinery liquid residual. Figure 4 show the plots of the CO behavior as a function of temperature. As is shown in Figure 4, the maximum CO value was presented about 400 ºC and 450 ºC.

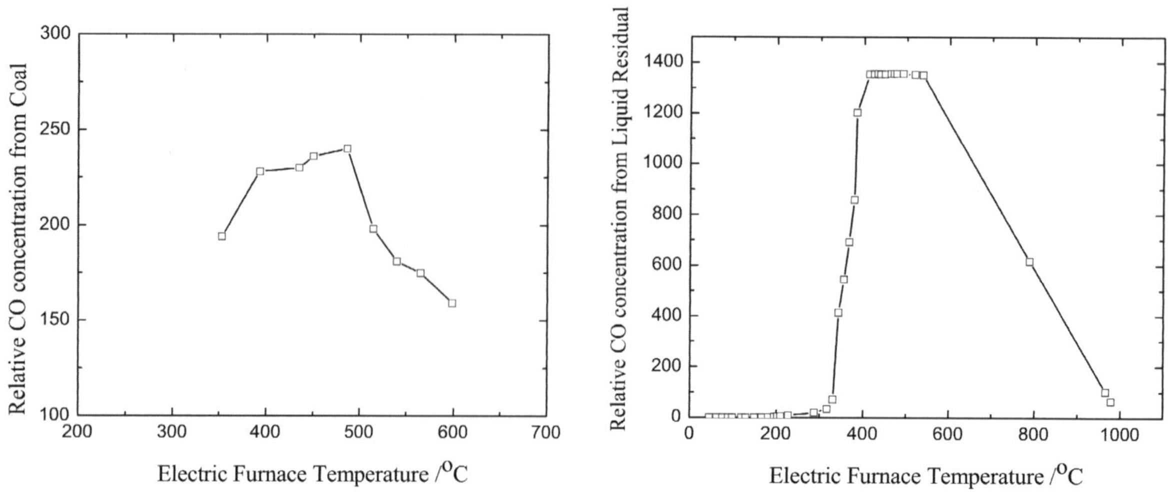

(a) (b)

Figure 4. CO behavior as a function of temperature for (a) Micare coal, and (b) Tula refinery liquid residual.

There was performed several experimental runs with small differences in the set up, or material used with similar results as the presented previously. As a result for these preliminary experiments, a gasification system at work-bench scale for a continuous process was designed. The design is described briefly in the next section.

IV. DESIGN OF A CONTINUOUS GASIFICATION SYSTEM AT WORK-BENCH SCALE

The gasification system proposed was in base of the electrical furnace used in the preliminary experiments. Since it is going to operate at atmospheric pressure, there is not need to special unique container. The gasification system should be modular.

Each module design is a replicate of the original electric furnace with some differences, i.e. each module has a temperature control independently, therefore the system can operate at the same temperature or to have a temperature profile along the system. Figure 5 shows the first module of the gasification system and some operational testing.

The high proposed of the completed gasification system is 1.50 m. Hence, there should be 5 modules of 30 cm each. Inside of each module, the electrical resistance can reach to up 1375 ºC and it is protected by a ceramic furnace process tube.

Actually, there are constructed six modules with a protection support, as shown in Figure 5. Also, the IIE lab is under the procedure of acquisition of complementary equipment to finish the construction of the gasification system.

Figure 5. First module of the gasification system design at the IIE.

It is planned to perform experimental runs of gasification of the same coal and liquid residual in continuous process. Also, a variety of Mexican refinery residuals i.e. liquid and solid as the petcoke, and some imported coal used already in some of the Thermoelectrical Power Plants are projected to be studied in the gasification system. The results will be presented in the future.

V. CONCLUSIONS AND FUTURE WORK

It is well known that the gasification process focused especially on the power generation is now under study around the world. The IIE has an ambitious project to study and to gather knowledge of the gasification process and the complementary process needed for an IGCC power plant.

The development and construction of the modular gasification system at atmospheric pressure will allow us to gather information of the kinetics gasification process, hence a gasification reactor design can be achieve for a pressurized system. Also a CFD modeling can be performed simultaneously on the basis of the Mexican feedstock available to use in IGCC power plant.

Actually, there are some design systems at different scale waiting to be approve for its construction and testing. Obviously, the system designs will be modified on the basis of the experimental results obtained on the modular gasification system discussed earlier.

More studies and research on the gasification technology and all its complementary systems, i.e. a separation air unit, feedstock systems, clean syngas system, CO_2

sequestering and storage system, syngas turbine systems, etc., is required for the design of a IGCC power plant.

There are several operational and physical-chemistry parameters from the fuel characterization that affect the syngas production in a gasification process. The work reported here is just a begging of a systematic experimental study to understand the black box represented by the gasifier reactor.

The understanding of the chemical reactions, chemical kinetics, and both mass and energy balances can favoured the syngas prediction with much real approach to the experimental gasification results. Also, CFD simulations are needed with different operational parameters conditions to short the experimental tests.

REFERENCES

ASTM (2005), "Standard classification of coal by rank", Designation: D388-05.

De Souza-Santos, Marcio L.; (2004) "Solid Fuels Combustion and Gasification. Modeling, Simulation, and Equipment Operation", Marcel Dekker, Inc..

Fernández-Montiel, M. y Alanís, F.J. (1993) "Reporte del Estado Actual de las Tecnologias de Gasificacion Aplicadas a la Generacion de Energia Electrica", Informe 44207, Instituto de Investigaciones Eléctricas, Octubre 1993, México.

Fernández-Montiel, M. y Alanís, F.J. (1994), "Estudios de Factibilidad para Plantas de Gasificacion de Carbon y de Combustoleo Integradas a Ciclos Combinados", Informe 44288, Instituto de Investigaciones Eléctricas, Marzo 1994, México.

Fernández-Montiel, M. (1994), "Reporte de actualizacion sobre el estado que guardan las tecnologias de gasificacion aplicadas a la generacion de energia electrica", Informe 44256, Instituto de Investigaciones Eléctricas, Octubre 1994, Mexico.

Fernández-Montiel, M., Alcaraz-Calderón, A. M. y García-Chávez, L. (2003), "Estudios y Desarrollo de Alternativas de Gasificación para el Sector Eléctrico", Informe 12585, Instituto de Investigaciones Eléctricas, Diciembre 2003, México.

Higman C. and Der Burgt M. Van, (2003) "Gasification", Elsevier.

Manzanares-Papayanopoulos*, E., Fernández-Montiel, M., Altamirano-Bedolla, J.A., Alcaraz-Calderón, A.M., and Romo-Millares, C.A., "Development and Evaluation of the Gasification Technology as Alternative Power Generation", AIChE Spring Meeting 2006, Paper 128e, April, 2006, Orlando, Fl., USA.

Rezaiyan J. and Cheremisinoff, N. P. (2005), "Gasification Technologies: A Primer for Engineers and Scientists", Taylor & Francis.

SENER (Secretaría de Energía) (2006), Prospectiva del Sector Eléctrico 2006-2015, Dirección General de Planeación Energética, Primera Edición.

Factors Limiting the Electron Transfer in Methanol Dehydrogenase Enzymatic Fuel Cells

Katherine L. Keeton, Rajeswari Veeramallu, and Daniela S. Mainardi. Louisiana Tech University, 911 Hergot Ave., Ruston, LA 71272

Recent advances in nano/bio technology are eliminating the past well-known power output limitations that bio-fuel cells (micro-Watts/cm2) faced with respect to chemical fuel cells (Watts to milli-Watts/cm2). Enzymatic fuel cells are currently being used to make electricity to power any number of electrical devices, such as pumps, valves and pacemakers, or electronic devices including but not limited to radios, sensors, controllers, and processors.[1] Currently a biocatalytic fuel cell that uses bacterial Methanol Dehydrogenase (MDH) enzyme as the anodic catalyst immobilized on N,N,N',N'-tetramethyl-p-phenylenediamine (TPMD)-functionalized carbon paste electrode, produces a continuous power output of 0.25 mW/cm2 for 30 days of continuous operation,[2] although the expected (theoretical) output is 100 mW/cm2.[3] It is believed that the TPMD mediator, responsible for electron transfer resulting from the fuel oxidation by the enzyme to the electrode, is the major factor limiting the power output, and deteriorating the fuel cell performance. In this work interaction between the MDH enzyme and the mediator TPMD is investigated using diverse molecular modeling techniques. Insight into how these molecules network to facilitate electron transport is achieved using quantum mechanical Density Functional Theory (DFT) and Molecular Dynamics simulations to obtain geometry, electronic configurations, and binding energies. The structure and stability of MDH enzymes and MDH/mediator complexes immobilized on electrode surfaces is investigated to further understand their behavior upon methanol oxidation. Moreover, studies of the enzyme conformational changes upon immobilization on electrode supports will be conducted and their effects on the enzyme activity and effectiveness for methanol oxidation elucidated.

1. Karyakin, A.A., S.V. Morozov, E.E. Karyakina, V. S.D., N.A. Zorin, and S. Cosnier, Hydrogen Fuel Electrode Based on Bioelectrocatalysis by the Enzyme Hydrogenase. Electrochem. Comm., 2002. 4: p. 417-420. 2. Zhang, X.-C., A. Ranta, and A. Halme, Direct methanol biocatalytic fuel cell-Considerations of restraints on electron transfer. Biosensors and Bioelectronics, 2006(21): p. 2052-2057. 3. Zhang, X. C., A. Ranta, and A. Halme. Effect of Different Catalytic Oxidants on the Performance of a Biocatalytic Methanol Fuel Cell. in Proceedings 204th Meeting of The Electrochemical Society. 2003.

Adsorptive Removal of Aromatic Compounds from Fuel Produced by Biomass Gasification for Fuel Cells

Yoshinori Itaya[1], Motoya Sakabe[1], Jun Kobayashi[2], Nobusuke Kobayashi[1], Shigenobu Hatano[1], and Shigekatsu Mori[3]. (1) Chemical Engineering, Nagoya University, Furo-cho Chikusa-ku, Nagoya, 464-8603, Japan, (2) Center for Cooperative Research in Advanced Science & Technology, Nagoya University, Furo-cho Chikusa-ku, Nagoya, 464-8603, Japan, (3) Emeritus Professor, Nagoya University, Furo-cho Chikusa-ku, Nagoya, 464-8603, Japan

Organic wastes such as woody biomass are converted into the fuel gas by the methane fermentation or the high temperature gasification. The research of the waste gasification processes is paid attention to produce fuels for fuel cells. The fuel gas generated in these processes contains generally unsaturated hydrocarbon such as aromatic compounds in the order of hundreds ppm. The existence of such compounds causes the carbon deposition on electrode catalysis in fuel cells or reforming of the fuel. Therefore, the process to remove those compounds from the gasification gas becomes essentially necessary. In this work, the adsorption removal process using the activated carbons was proposed. The benzene adsorption characteristic of various activated carbons was evaluated, and the adsorption removal experiment under an actual gasification gas composition was carried out. The activated carbons made from the coconut shell, coal and the phenolic resin were employed for this examination. The adsorption experiment with the thermogravimetric analysis was carried out to clarify the adsorption characteristics of benzene as a representative sample of aromatic compounds on the various activated carbons. The characteristics of benzene adsorption on the activated carbons were evaluated at 303 K of adsorption temperature and 500 ppm of initial concentration. For preparation of the activated carbons, degassing of them had been carried out at 473 K for 60 minutes. From the results of equilibrium amount of adsorbed benzene, it was found that the activated carbons made from the coconut shell and the phenolic resin had same level of high adsorption ability. However, the activated carbon made from the phenolic resin is too expensive to apply to this removal process. Therefore, the activated carbon made from the coconut shell was used to the following investigation. To obtain adsorption isotherm of the activated carbon, equilibrium amount of adsorbed benzene at 303 K was measured in the range of 100-1000 ppm. The adsorption isotherm could be expressed as the Freundlich isotherm equation, such as $q = 1.45 \mathrm{x} C^{0.235}$. Here, q is the amount of adsorbed benzene [kg/kg] and C is the concentration of benzene [-]. Using fixed-bed adsorber of the activated carbon, breakthrough of benzene adsorption was measured. Simulated gasification gas, of which composition was H2: 15%, CH4: 3%, CO: 25%, CO2: 25%, benzene: 200 ppm and N2 balance, was supplied into the fixed-bed adsorber at 1.5 *l*/min and transient behavior of benzene concentration at the outlet of the bed was measured. Adsorption temperature was 303 K and the height of fixed-bed was about 100 mm in a cylindrical glass tube of 22.5 mm in diameter. The benzene concentration in the exit gas was maintained at about 4 ppm for 4 hours or more. It was indicated that removal of benzene from the gasification gas was possible even if other gases such as CO and CO2 coexistent with benzene.

Molten Carbonate Fuel Cell Hybrid Power System Simulation

Brian Wolf and Shripad T. Revankar
School of Nuclear Engineering
Purdue University
West Lafayette, IN 47907
Email: wolfb@purdue.edu

ABSTRACT

Advancement in research of distributed generation of electrical power is a result of energy security issues and changing markets and technologies. Fuel cell hybrid technologies which integrate high temperature fuel cells with another power generation technology have promising abilities which make them an important research topic for development and commercialization. A hybrid system consisting of molten carbonate fuel cells (MCFC) and gas turbines has been considered as a distributed energy system. Steady state and dynamic models of the system were developed. The model was implemented using SIMULINK software and simulation of a plant trip, in which the fuel cell load starts at steady state and abruptly goes to zero has been performed. The steady state and dynamic behavior has been benchmarked with model results in literature. The model uses differential equations to calculate mass fractions, pressure, and temperature with basic simplifying assumptions.

NOMENCLATURE

M	mass
C_p	Specific heat capacity at constant pressure
T	Temperature
F	Total molar flowrate
y	Mole Fraction
$\tilde{C}_t$	Average total molar concentration
$\bar{h}$	molar enthalpy
R	Rate of production
ν	Stoichiometric coefficient of species in reaction
Q_l	Convective heat loss
r	Reaction rate
k	Reaction rate constant
K	Equilibrium constant
P	Pressure of Power

Superscripts

o	anode (cathode) exit
S	Fuel cell stack
in	inlet

Subscripts

DC	DC electrical power
a	Anode Side
c	Cathode side
i	Species index
μ	Total number of reactions
ξ	Total number of species
j	Reaction index

INTRODUCTION

In the last few years, there has been much interest in distributed generation. While the first electric power systems were in themselves distributed generation systems used to meet the demand of small communities, they were driven by economics to become the large centralized systems we see today. Research and development since the beginning of power were focused on spreading the grid over entire regions, with increasing loads and demands. While economic, the current grid system may not meet the needs of future technology which requires stable and reliable power. Also, large centralized power systems are hard to secure and are vulnerable to large power outages.

Fuel cell systems offer the ability to produce power at very high efficiency with little or no pollution emissions, allowing for future energy demands to be met as well as setting new standards for emissions criteria. Fuel cell systems are receiving much needed attention in both research and development due to their promising characteristics for being the next generation power plant technology. Much progress has been made in developing hybrid systems for distributive generation.

A hybrid system can be characterized as having an efficiency greater than the sum of it's individual components efficiencies. Hybrid systems, which are typically a high temperature fuel cell coupled with a turbine, have been studied and show remarkable efficiencies of over 65 percent [1-10]. Chan et. al. developed a 1.3 MW solid oxide fuel cell and gas turbine system with an overall system efficiency of over 80% [2]. Steinfeld et. al. showed simulation studies of a 200 MW molten carbonate fuel cell and gas turbine plant producing an efficiency of over 70%. This model assumed that high temperature heat exchanger technology currently being developed is available[6]. Although technical hurdles still exist in the development of such systems,molten carbonate fuel cells (MCFCs) and solid oxide fuel cells (SOFCs) are now being tested in pilot power plants in the US and other countries. These high temperature fuel cells make heat available to be recovered and used in a secondary system such as a turbine cycle. Their promising abilities make hybrid systems which incorporate high temperature molten carbonate or solid oxide fuel cells an important research topic for development and commercialization of combined heat and power systems. They meet many demands of U.S. energy goals for the future, including independence from foreign sources, greater security, and pollution free emissions. Though impressive, they still require a great deal of development effort if they are to be commercialized.

HIGH TEMPERATURE FUEL CELLS

High temperature fuel cells offer the ability to use natural gas as the fuel. In this case the methane based fuel must be reformed in order to produce the hydrogen needed for the electrochemical reaction. In the case of a MCFC there are 4 main chemical reactions that must take place as described in equations (1)-(4). Equations (1)-(3) describe reactions on the anode side, while (4) describes the cathode side [11].

Anode Side:

$$H_2 + CO_3^{=} \rightarrow H_2O + CO_2 + 2e^- \quad \text{Anode} \tag{1}$$

$$CO + H_2O \rightarrow CO_2 + H_2 \quad \text{Water-Gas Shift} \tag{2}$$

$$CH_4 + H_2O \rightarrow CO + 3H_2 \quad \text{Reforming} \tag{3}$$

Cathode Side:

$$\frac{1}{2}O_2 + CO_2 + 2e^- \rightarrow CO_3^{=} \quad \text{Cathode} \tag{4}$$

SOFC's follow a similar reaction scheme however for simplicity only MCFC's will be discussed in this paper. It is important to understand that the reactions (1) and (4) are electrochemical reactions and their rates are proportional to load current described by Faraday's Law. The water-gas shift and reforming reactions however are dependent on many things such as catalyst loading, temperature, pressure and gas composition [12]. The reaction rates of the reforming and water-gas shift reactions favor high temperatures and low pressures. Hybrid systems rely on the fact that the electrochemical reactions are highly exothermic. While the water-gas shift reaction is mildly exothermic, the reforming reaction is highly endothermic. The electrochemical reaction however produces enough heat to sustain the reforming reaction, making the overall fuel cell exothermic. There are 4 main designs of high temperature fuel cells each of which handles the reforming and water-gas shift reactions differently. The following figure helps describe the differences. In Fig. 1 (a) an external reforming (ER) unit is used to reform the methane to create a hydrogen rich fuel before it enters the anode. It would be more advantageous to use the heat from the fuel cell to aid in reforming which increases the efficiency of the overall system. As seen in Fig.1 (b) an indirect internal reformer (IIR) is used. The IIR consists of a catalyst bed which reforms approximately 50-60% of the natural gas to hydrogen before it enters the anode. The IIR is in close contact with the fuel cell allowing utilization of the heat released from the fuel cell reactions. A similar method is to use a direct internal reformer (DIR) which allows the reforming reactions to take place directly at the anode compartment. A more recent method is to use a combination of both IIR and DIR as seen in Fig. 1 (d). Here the fuel is partially reformed in the IIR and then enters the DIR where it is further reformed allowing greater utilization of the methane fuel [11].

HYBRID SYSTEMS

Incorporating a gas turbine (GT) with a high temperature fuel cell further increases power production as well as system efficiency. There are many plant configurations that have been analyzed [1-10]. For example, Fig. 2 shows a basic configuration of an MCFC-GT power plant. Waste heat from the exhaust of the fuel cell is exchanged to air, the working fluid of the compressor-turbine combination. This is then used to oxidize anode exhaust in order to utilize the existing CO and CO2 needed in the cathode for electrochemical reactions. The added cycle introduces complexity to the system requiring extensive thermodynamic analysis in order to optimize the plant. This layout is just one of many possible options that may be considered. The layout will have a great affect on how well the system will perform on a thermodynamic scale, so more than one layout shall be analyzed. Another key feature of hybrid systems is their relatively easy ability to be scaled up or down. Fuel cells are arranged in stacks in both series and parallel in order to increase power output. By changing the size of the fuel cell system, plants can be designed to accommodate demands from a few kW to multi-MW. Typically, the fuel cell produces approximately 70% of the total power of the system, while the turbine produces about 30%. Although these plants use natural gas as a fuel, their increased efficiency allows for a higher power output for the same amount of fuel compared to current generation methods. This reduces the amount of emissions produced in power production.

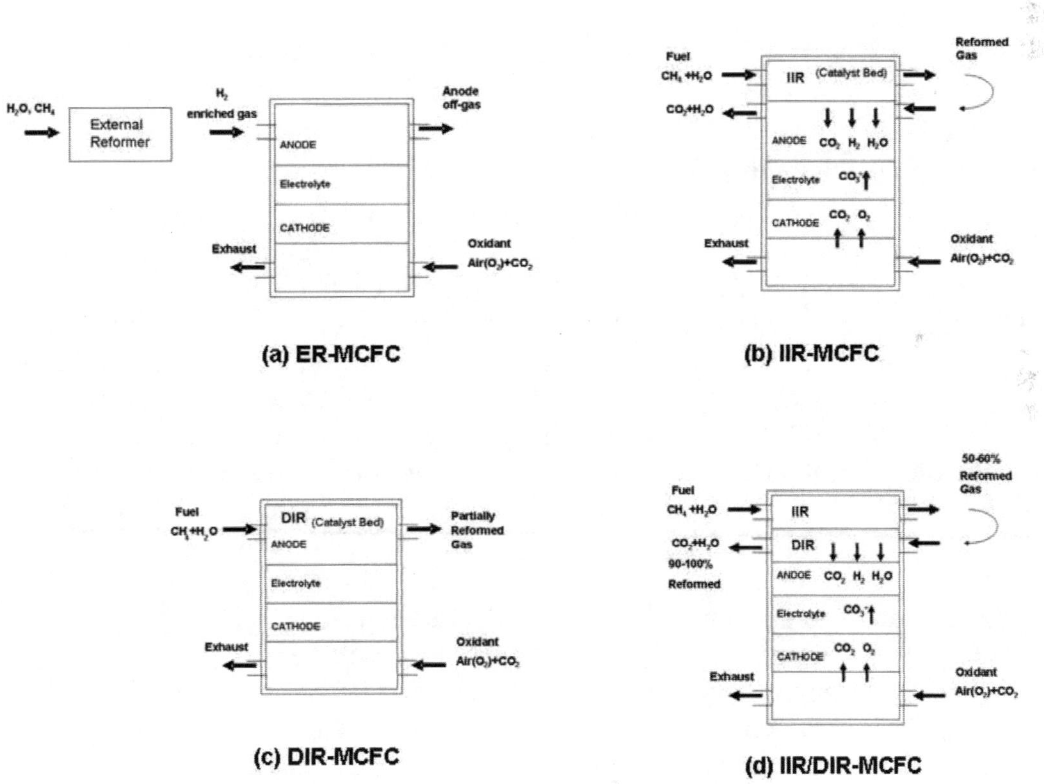

Figure 1: Schematic representation of placement for reforming in MCFC

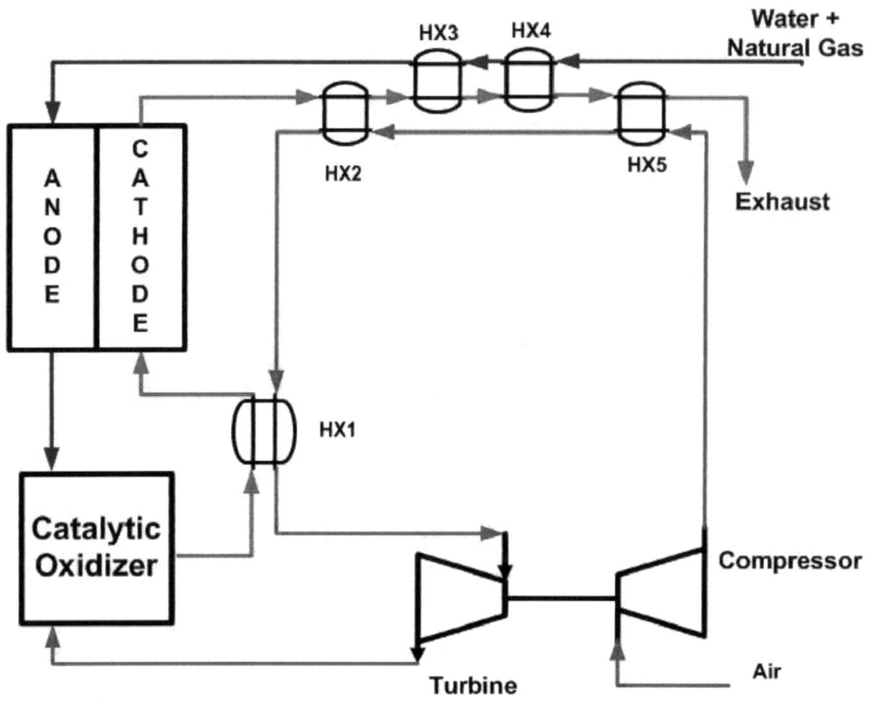

Figure 2: Example hybrid system layout

DISTRIBUTED GENERATION

The hybrid systems as described above can be used in many applications. Hospitals, water pumping and treatment facilities as well as computer systems currently have large generators or batteries to supply standby power when needed, which could be replaced by such hybrid technology. Also, current power utilities and customers would benefit by using distributive generation by utilizing on-site power during peak periods which can be costly as well as put strain on the grid. Distributive generation could also accommodate areas where the power grid has problems reaching or delivering power. Also, distributive generation can be used where transmission and distribution grid service is not available.

MODELING

The key goal is to develop a dynamic model of a hybrid system including controls and a state of the art transmission grid. Each component will be modeled separately and then joined with a common software. One of the key questions at hand is how each component acts by itself compared when integrated into the hybrid system. Also, transient response is important in order to determine the role each component will play in a load increase or decrease and to find the limitations of the system. The modeling tasks are part of a collaborative project with Wright State University, where the compressor and turbine models are being developed. The remainder of this paper will focus only on the fuel cell model development.

FUEL CELL MODEL

There is much to gain by incorporating the fuel reforming unit within the fuel cell as the IIR/DIR type explained previously. A few advantages of such a system are the higher methane conversion as well as a lower system cost. Also, because the reforming reaction takes advantage of the heat released by the electrochemical reactions, there is less heat to be removed and subsequently lost by convection from the fuel cell. Not only does this raise overall efficiency of the system, but it has a positive effect on the temperature distribution of the fuel cell by lowering the temperature difference across the stack. For this reason, an IIR/DIR fuel cell is an excellent candidate for future hybrid systems and has been chosen for modeling in this study. Lukas et. al. [13] have developed an explicit dynamic lumped parameter model of such a system rated at 2 MW DC power. The model treats the anode and cathode as a continuosly stirred tank reactor by lumping the indirect and direct reforming units. This method has been shown to correlate well with experimental data as described by Lukas [14].

An energy balance considers a single temperature common to both the anode and cathode and only accounts for energy storage with in the materials making up the stack hardware as seen in equation (5). Energy storage in the gas species residing in the stack is small compared to that of the stack hardware and therefore is neglected. Also, all species are considered to be ideal gases and thier exit temperature is equal to the stack temperature.

$$M^s C_p^s \frac{dT^s}{dt} = F_a^{in}[\sum_{i=1}^{\xi} y_{ai}^{in}(\bar{h}_{ai}^{in} - \bar{h}_{ai}^{s})] - \sum_{i=1}^{\xi} \bar{h}_i^s R_{ai} + F_c^{in}[\sum_{i=1}^{\xi} y_{ci}^{in}(\bar{h}_{ci}^{in} - \bar{h}_{ci}^{s})] - \sum_{i=1}^{\xi} \bar{h}_i^s R_{ci} + Q_l - P_{DC} \qquad (5)$$

The first and third terms on the right hand side of equation (5) is the energy change due to flow into and out of the anode and cathode based on inlet molar flowrate. These terms do not account for changes due to reactions therefore, the second and fourth terms correspond to the heat produced or lost by reaction in both anode and cathode. The rates of production are defined in equation (6). The last two terms, Q_l and P_{DC} represent convection heat loss and power losses.

$$R_i = \sum_{j=1}^{\mu} v_{ij} r_j \qquad (6)$$

A general molar balance, one for the cathode and one for the anode, can be written as

$$V\tilde{C}_t \frac{dy_i}{dt} = N^{in}(y_i^{in} - y_i^{o}) - y_i^{o} \sum_{i=1}^{\xi} R_i + R_i, \qquad i = 1, \ldots, \xi, .. \qquad (7)$$

Since the electochemical reactions rates of (1) and (4) are proportional to load current through Faraday's law, they are rather easily determined. The reforming reaction and water-gas shift reaction however, are dependent on many parameters and if defined seperately could add much complexity to the system model. It is therefore important to simplify the reaction kinetics as much as possible. With this in mind the water-gas shift reaction is assumed to be in equilibrium due to the fact that it has a very fast reaction rate compared to that of the reforming reaction. This imposes a simple constraint on gas species that appear on the anode side described below and allows its reaction rate to be left undefined [12].

$$K_{wgs} = \frac{y^s_{CO_2} \cdot y^s_{H_2}}{y^s_{CO} \cdot y^s_{H_2O}} \tag{8}$$

Here, K_{wgs} is the water-gas shift reactions equillibrium constant dependant on Gibb's free energy and stack temperature.
This allows reactions (2) and (3) to be lumped into an equivalent reaction (9).

$$CH_4 + 2H_2O \rightarrow CO_2 + 4H_2 \tag{9}$$

This reaction has been studied extensively and in 1975 Rostrup-Nielsen [15] proposed a simple expression to describe its reaction rate below.

$$r_{ref} = kP_{CH_4}[(1-Q)/K_{ref}] \qquad (mol/s) \tag{10}$$

Here k is a rate constant and Q is the mass action expression dependent on partial pressures. K_{ref} is the reforming rate equilibrium constant. Applying these equations as well as others to the lumped parameter model described by Lukas et. al. [12,13] a fuel cell model was developed to describe the behavior of a smaller 250 kW fuel cell made up of two 125 kW fuel cells in series.

Table 1: Steady State values of 250 kW fuel cell

Current (mA/cm^2)	160			
DC Power (kW)	250			
Mole Fraction	Anode Inlet	Anode Outlet	Cathode Inlet	Cathode Outlet
H_2	0.1168	0.0741	0.0000	0.0000
CH_4	0.2798	0.0011	0.0000	0.0000
CO	0.0005	0.0460	0.0000	0.0000
CO_2	0.0346	0.4535	0.1553	0.0476
H_2O	0.5662	0.4244	0.1553	0.1880
N_2	0.0000	0.0000	0.5599	0.6778
O_2	0.0000	0.0000	0.1294	0.0866
Temperature (K)	849	950	838	950
F (mol/s)	2.203	4.463	17.62	14.56

STEADY STATE

The equations above along with many others were incorporated into a Matlab/SIMULINK code. The model consists of ten differential state equations and over 150 algebraic calculations. Table 1 shows steady state values of the system. Inlet and outlet mole fractions are listed corresponding to each species. In this study, inlet properties were held constant. There were no external controls to the system except the steam to carbon ratio is set equal to 2 at the anode inlet. Also, 75% fuel utilization was assumed and also held constant to determine the flowrate into the anode. The steady state results correspond to a consant full load of 250 kW DC power. It is of interest that 99% of the methane has been reformed in the anode. Also, the electrochemical reactions raised the stack temperature 100 K above inlet temperatures.

TRANSIENT RESPONSE

The transient response of the fuel cell to a plant trip, where the full load current of 160 mA/cm^2 is instantaneously removed during steady state operation is shown in Figs. 3 and 4. During the transient the anode and cathode inlet properties are held constant. Fig. 3(a) shows the the sudden affect on the gas composition when the electrochemical reaction rate at the anode decreases to zero. The fuel cell consumes less hydrogen while the reforming reaction keeps producing hydrogen at nearly the same rate. Fig. 3(b) is shown to analyze the long term affects of the relatively slow reforming reaction rate in the anode compartment. This slow reforming rate can be explained by studying the very slow response of the stack temperature in Fig. 4(c). This slow response is due to the ability of the stack to store massive amounts of energy because of its very large mass-specific heat product. This is an excellent quality because during a transient such as this, heat can still be generated to maintain the turbine side power. It can be seen in Fig. 3(c) that the composition in the cathode responds very quickly because only electrochemical reactions take place, therefore no slow response occurs. Due to the dependance of the cell voltage on both anode and cathode compositions as well as stack temperature, there exists both fast and slow time behavior as seen in Figs. 4(a) and 4(b).

CONCLUSION

This paper explains distributed generation as well as the concepts behind internal reforming of MCFCs. Assumptions which simplify modeling of fuel cells were mentioned as well as basic balance equations. Steady state results are shown and represent expected values. A power plant trip accident was used to explain the transient response of the fuel cell on both short and long time scales. As mentioned previously, future work inlcudes the coupling of the fuel cell with a turbine/compressor combination as well as balance-of-plant. Heat recovery equipement as well as a catalytic burner and controls will be implemented in order to obtain an optimized system. Systems on both small (250 kW) and large (2 MW) scales will be discussed.

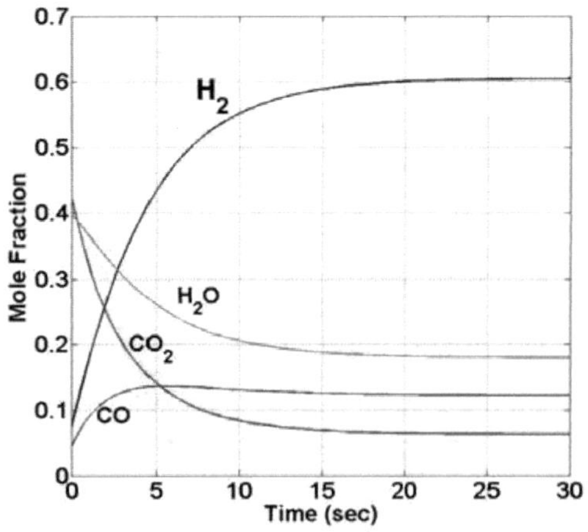

(a) anode, short response

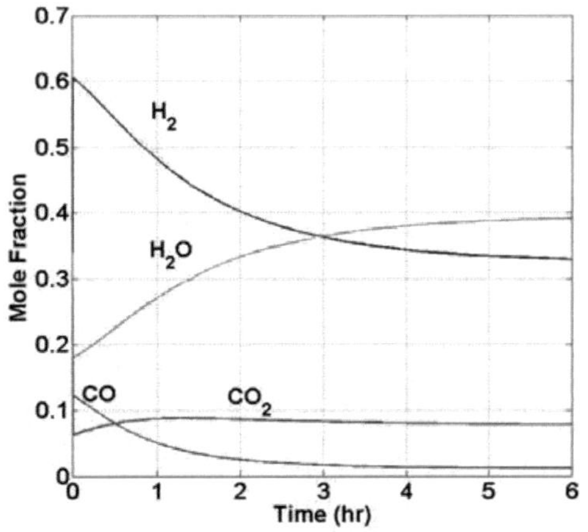

(b) anode, long response

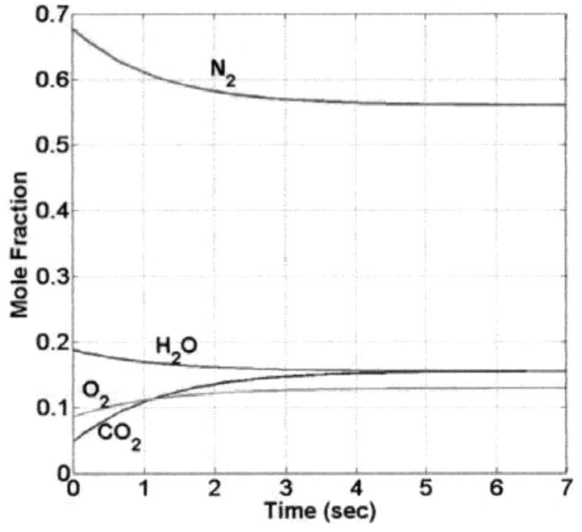

(c) cathode

Figure 3: Anode and cathode exit compositions

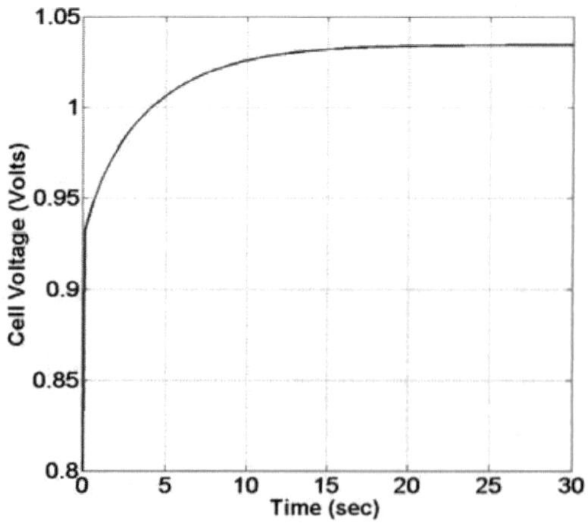

(a) short time voltage response

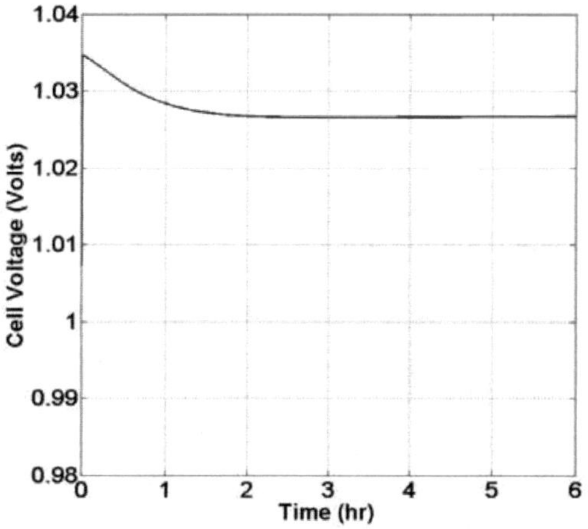

(b) long time voltage response

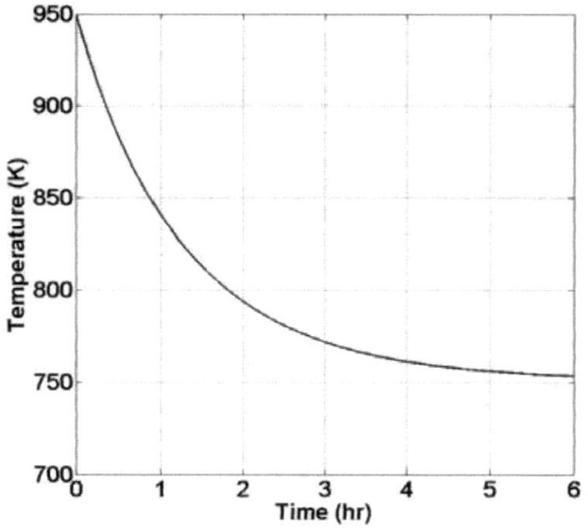

(c) stack temperature

Figure 4: Cell voltage and stack temperature

REFERENCES

[1] Amorelli A, Wilkinson MB, Bedont P, Capobianco P, Marcenaro B, Parodi F, Torazza A: An experimental investigation into the use of molten carbonate fuel cells to capture CO2 from gas turbine exhaust gases. Energy, Genova, Italy 2004, 29:1279-1284.

[2] Chan SH, Ho HK, Tian Y: Multi-level modeling of SOFC-gas turbine hybrid system. International Journal of Hydrogen Energy 2002, 29:889-900.

[3] Chan SH, Low CF, Ding OL: Energy and exergy analysis of simle solid-oxide fuel-cell power systems. Journal of Power Sources 2001, 103:188-200.

[4] Farooque M, Maru HC, Baker B: Direct carbonate fuel cell power plant design at ERC. In Proceedings of the 28th intersociety energy conversion engineering conference; Atlanta, GA, USA: 1993:75-78.

[5] Fogler SH: Elements of Chemical Reaction Engineering edn 3rd. New Jersey: Prentice Hall; 1999.

[6] Ghezel-Ayagh H: Direct Fuel Cell/Turbine Power Plant. Fuel Cell Annual Report 2003:115-118.

[7] Grillo, Magistri, Massardo: Hybrid systems for distributed power generation based on pressurization and heat recovering of an existing 100 kW molten carbonate fuel cell. Journal of Power Sources 2002, 115:252-267.

[8] Hirshenhofer JH, Stauffer DB, Engleman RR, Klett MG: Fuel Cell Handbook; 1998.

[9] Jurado F: Study of molten carbonate fuel cell-microturbine hybrid power cycles. Journal of Power Sources 2002, 111:121-129.

[10] Kimijima S, Kasagi N: Cycle Analysis of Micro Gas Turbine-Molten Carbonate Fuel Cell Hybrid System. Trans. Jpn. Soc. Mech. Eng. 2004, 48:1001-1008.

[11] Larminie J, Dicks A: Fuel Cell Systems Explained. New York, NY: John Wiley and Sons Inc.; 2003.

[12] Li X: Principles of Fuel Cells edn 1. New york: Taylor and Francis Grp.; 2006.

[13] Lukas MD, Ghezel-Ayagh H, Cervi MC: Experimental Transient Validation of a Direct FuelCell Stack Model. Proc. of the 2001 of the IEEE Power Engineering Soc. Summer Meeting, Vancouver 2001.

[14] Lukas MD, Ghezel-Ayagh H, Cervi MC: Experimental Transient Validation of a Direct FuelCell Stack Model. Power engineering Society Summer Meeting IEEE 2001, 3:1363-1368.

[15] Lukas MD, Lee KY, Ghezel-Ayagh H: Development of stack simulation model for control study on direct reforming molten carbonate fuel cell power plant. IEEE Trans. Energy Conversion 1999, 14:1651-1657.

[16] Lukas MD, Lee KY, Ghezel-Ayagh H: An Explicit Dynamic Model for Direct Reforming Carbonate Fuel Cell Stack. IEEE Trans. Energy Conversion 2001, 16:289-295.

[17] Rostrup-Nielsen JR: Steam Reforming Catalysts. Danish Technical Press Inc., Copenhagen 1975:85.

[18] Steinfeld G, Maru HC, Sanderson RA: High Efficiency Carbonate Fuel Cell/Turbine Hybrid Power Cycle. IECEC conference paper 1996.

[19] Wolf TL, Wilemski G: Motlen Carbonate Fuel Cell Performance Model. Journal of Electrochem. Soc. 1983, 130:48.

[20] www.netl.doe.gov: Hybrid Power Systems to Achieve High Efficiency. U.S.D.O.E.

[21] Yi Y, Rao AD, Brouwer J, Samuelsen GS: Analysis and optimization of a solid oxide fuel cell and intercooled gas turbine (SOFC-ICGT) hybrid cycle. Journal of Power Sources 2004, 132:77-85.

Study of Gas Diffusion Layers in Direct Methanol Fuel Cells

Virendra K. Mathur[1], Jason Morgan[1], and William Grant[2]. (1) Dept. of Chem. Eng., University of New Hampshire, 260 Kingsbury Hall, Durham, NH 03824, (2) New Products, Ballard Material Products, 2 Industrial Ave., Lowell, MA 01851

Fuel cells are electrochemical devices that convert chemical energy of a reactor fuel directly into electricity with high efficiency and a minimum of pollution. Proton Exchange Membrane (PEM) fuel cells are among one of the most promising energy sources for transportation and small appliance applications operated on methanol as a fuel. Some of the key materials used in PEM fuel cells are catalysts, membranes, gas diffusion layers (GDLs) and bipolar plates. There is a need for the development of GDLs which can be tailored for continuous commercial scale production of membrane electrode assemblies (MEAs) to reduce the labor costs of fuel cell manufacture. This study is being conducted for the development of paper and fabric GDLs for the use in the Direct Methanol Fuel cells. DMFC (5 cm2 single cell) operating on methanol and air has been used in this investigation. Membrane used is Nafion 115. The effect of experimental parameters such as cell temperature, methanol concentration and feed rate, catalyst loading etc. have been studied to maximize the cell performance. Comparative studies with other commercially available non-woven and woven GDLs are also presented.

Development of Direct Alkaline Fuel Cell for Liquid Fuels

Suddhasatwa Basu, Chemical Engineering, IIT Delhi, Hauz Khas, New Delhi, India and A. Verma, Chemical Engineering, IIT Guwahati, Guwahati, India.

Extent of energy generation and its use is the backbone of growth for any country. The problems of limited resources and environmental pollution by fossil fuel have given thrust to the search for the alternative energy generating methods using renewable resources. Fuel cell technology is found very promising as an alternative method for converting chemical energy into electrical energy using hydrogen or hydrogen rich organic and inorganic compounds. Towards this approach, an alkaline fuel cell with potassium hydroxide solution as an electrolyte is developed for the direct use of methanol, ethanol or sodium borohydride as fuel.

Electrodes for the fuel cell were prepared using Pt-black, which acted as anode and a non-noble metal catalyst, MnO2, which acted as cathode. Nickel mesh was used as current collector. The electrodes were characterized by scanning electron micrography and cyclic voltammetry analyses and then used to fabricate the direct methanol or ethanol or sodium borohydride alkaline fuel cell. The current density-cell voltage characteristics of the cell was determined by measuring voltage and current at different experimental conditions e.g., electrolyte concentration, fuel concentration and temperature. The highest performance at 25oC is obtained with 3M KOH electrolyte concentration and 3M methanol or ethanol and 2m sodium borohydride concentration. The maximum power density of 24.3 mW cm-2 and 14.5 mW cm-2 were obtained for methanol and ethanol at 65oC while that for NaBH4 was 22.5 mW cm-2 at 25oC.The life-time of the fuel cell is about 400 hours at 0.5 V and 22 mA cm-2 load conditions and it deteriorated at the rate of 25 ƒÝ V hr-1. The electrodes could be regenerated using HCl solution and more than 80% of the open circuit voltage was recovered.

Keywords: Alkaline fuel cell; Ethanol; Methanol; Lifetime of fuel cell; Sodium borohydride

Electro-Oxidation Study of Ethanol in Direct Ethanol Fuel Cell on Pt Based Catalyst

Suddhasatwa Basu, Chemical Engineering, IIT Delhi, Hauz Khas, New Delhi, India and H. Pramanik, Chemical Engineering, IT BHU, Benaras, India.

Direct ethanol fuel cell is the future power source of portable electronic equipments e.g., laptop, camcorder and mobile phone. Ethanol is attractive as compared to methanol because it is renewal in nature. Ethanol is produced through fermentation route from agricultural product. In view of this, a direct ethanol fuel cell (DEFC) is constructed. The anode of DEFC was made of PtRu (40%:20% by wt)/ C electrode-catalysts with variable loadings of 0.6-1.5 mg/cm2. The cathode was made of Pt-black and loading was fixed to 0.6 mg/cm2. Electro-oxidation of ethanol on PtRu was studied in half-cell mode using cyclic voltammetry technique. Polymer electrolyte membrane (PEM) was cast from Nafion® (SE-5112, Dupont) dispersion and the membrane electrode assembly was prepared by hot-pressing PEM in between anode and cathode at a given temperature and pressure. During the operation of the anode and cathode side temperature was varied whereas the pressure was maintained at 1 bar and the ethanol was fed at a flow rate of 1.2 ml/min. The cell voltage and current density was measured at different loads, temperatures and concentrations of ethanol and electrode-catalyst loadings. At anode temperature of 90oC and cathode 60oC for 2M ethanol with pure O2 supply at cathode, the maximum open circuit voltage (OCV) of 0.815 V, current density of 24.8 mA/cm2 and power density of 9.52 mW/cm2 were obtained at anode loading of 1 mg/cm2. For the same conditions but at an anode temperature of 42 oC and cathode 42 oC, the OCV of 0.810 V, current density of 22 mA/cm2 and power density of 7.33 mW/ cm2 were obtained. Cyclic volammetry data substantiates above anode conditions for obtaining of maximum current density. The power density obtained is lower than that reported in the literature for direct methanol fuel cell.

Keywords: direct ethanol fuel cell, half-cell, membrane electrode assembly, electrode-catalyst

Lithium/Air Battery Model

S.S. Sandhu[+,a], *J.P. Fellner*[b], *G.W. Brutchen*[a]

[a]*Department of Chemical and Materials Engineering, University of Dayton*
300 College Park, Dayton, OH 45469-0246
[b]*Air Force Research Laboratory, Propulsion Directorate, Plans and Analysis Branch, 1950 Fifth Street, Wright-Patterson AFB, OH 45433-7251*

Abstract

A diffusion-limited transient mathematical model for a lithium/air cell, with the air cathode pores flooded with an organic electrolyte, has been developed. During cell discharge, the cathode pore radius profile is reflective of the distribution of the lithium peroxide product in the cathode. The cathode pore radius profile has been predicted as a function of time, current density, oxygen gas pressure, and cathode thickness for an assumed initial porosity and average cathode pore size. Transient concentration profiles of the dissolved oxygen in the electrolyte were also determined. Capacities of the lithium/air cell were predicted and compared favorably with literature experimental results.

Keywords: lithium-air cell, transient model, organic electrolyte, diffusion-limited

[+]*Corresponding author: e-mail address: Sarwan.Sandhu@notes.udayton.edu*

1. Introduction

The formulation presented in this paper is for a lithium/air cell, consisting of a thin lithium sheet anode, lithium phosphorous oxynitride (LiPON) or glass anode protective layer, and a porous carbon oxygen/air cathode; see Fig. 1. The porous carbon cathode is assumed to be made of carbon black, catalyzed by a metal such as manganese, polytetrafluoroethylene (PTFE) as the binder, and the pores are flooded with a lithium based electrolyte, for example, 1 M $LiPF_6$ in PC/1,2-dimethoxyethane (DME) in a mass ratio of 1:1 (1:1).

2. Mathematical Formulation

Lithium ions are assumed to enter the anode lipon or glass protective layer at the interface between the lithium anode sheet and protective layer corresponding to a geometric current density value expressed by i_{geom}. The Li^+ ions migrate through the protective layer under the direct current electric field effect. The voltage loss across the layer is given by

$$\left|\Delta\Phi_{protect-l}\right| = \frac{i_{geom} l_{protect-l}}{k_{protect-l}} \tag{1}$$

where $k_{protect-l}$ is the lithium ion conductivity of the anode protective layer to protect the lithium anode sheet from water and carbon dioxide, and $l_{protect-l}$ the protective layer thickness. The

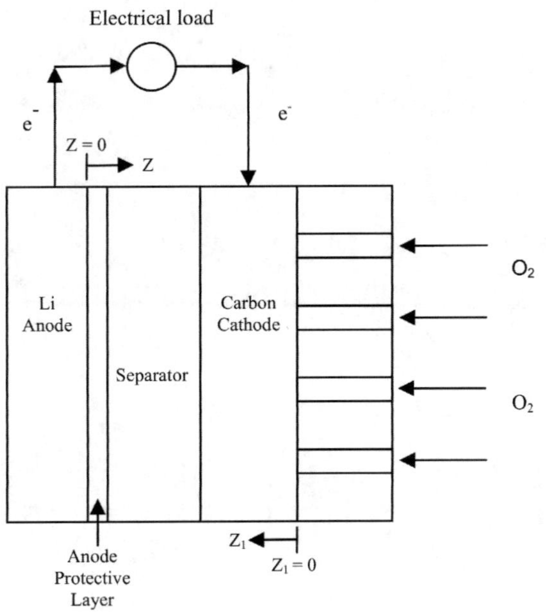

Figure 1. Simple sketch of lithium/air system.

lithium ion transport through the electrolyte, for example, $LiPF_6$, in an organic solvent held in polyacrylonitrile (PAN) layer as the separator between the anode sheet and cathode layer or in the pores of the carbon cathode layer in close contact with the anode protective layer, occurs under the effect of an ion activity gradient and a direct current electric field. The voltage gradient in the electrolyte is given [1] by

$$F\vec{\nabla}\Phi = -\frac{F\vec{i}_{geom}}{k_{electrol-eff}} - RT\sum_i \frac{t_i}{z_i}\vec{\nabla}\ln a_i \tag{2}$$

where Φ is the electric potential in the electrolyte, F the Faraday constant given by 96,487 Coulombs per g-equivalent or g-mol of H^+ ions or electrons, $k_{electrol-eff}$ the effective ionic conductivity of the electrolyte held in a cell layer, R the universal gas-law constant given by 8.314 J $mol^{-1}K^{-1}$, T the cell temperature, t_i the transfer number of an ion i, z_i the number of positive or negative charges on an ionic species, and a_i the ion activity of ion i in the electrolyte solution. The z-component of Eq. (2) is

$$F\frac{d\Phi}{dz} = -\frac{F}{k_{electrol-eff}} i_{geom} - RT\sum_i \frac{t_i}{z_i}\frac{d\ln a_i}{dz} \tag{3}$$

The integration of Eq. (3) over the electrolyte thickness, $l_{electrol}$, with respect to z leads to

$$F\left|\Delta\Phi_{electrol}\right| = \left(\frac{Fi_{geom}l_{electrol}}{k_{electrol-eff}}\right) + RT\sum_i \frac{t_i}{z_i}\ln\left(\frac{a^r_{i,electrol}}{a^l_{i,electrol}}\right) \tag{4}$$

where $\left|\Delta\Phi_{electrol}\right|$ is the magnitude of the voltage loss across the electrolyte region of thickness, $l_{electrol}$, and $a^r_{i,electrol}$ and $a^l_{i,electrol}$ are the ion activities on the right and left sides of the electrolyte region of thickness, $l_{electrol}$. For the electrolyte solution of $LiPF_6$ in an organic solvent, given that $z_+ = z_{Li^+} = +1$, $z_- = z_{PF_6^-} = -1$, and $t_{Li^+} + t_{PF_6^-} = 1$, Eq. (4) leads to

$$\left|\Delta\Phi_{electrol}\right| = \frac{i_{geom}l_{electrol}}{k_{electrol-eff}} + \left(\frac{RT}{F}\right)\left[t_{Li^+}\ln\left(\frac{a^r_{Li^+,electrol}a^r_{PF_6^-,electrol}}{a^l_{Li^+,electrol}a^l_{PF_6^-,electrol}}\right) - \ln\left(\frac{a^r_{PF_6^-,electrol}}{a^l_{PF_6^-,electrol}}\right)\right] \tag{5}$$

Activity of an ion, $a_{ion,electrol}$, is here expressed as

$$a_{ion,electrol} = f_{ion,electrol}\frac{c_{ion}}{c^\circ} \tag{6}$$

where $f_{ion,electrol}$ is the ion activity coefficient to account for the solution nonideal behavior, c_{ion} the molar concentration of an ion in the electrolyte solution, and c° = 1M (employed here to make $f_{ion,electrol}$ a dimensionless quantity). For the electrolyte, $LiPF_6 \rightleftharpoons Li^+ + PF_6^-$ (assuming complete dissociation), $\nu_+ = 1$, $\nu_- = 1$, $\nu_+ + \nu_- = 2$, and $c_{Li^+,electrol} = c_{PF_6^-,electrol} = c_{electrol}$ (salt molar concentration in the electrolyte solution). By definition, the salt activity coefficient, $f_{\pm,electrol}$, is related to the ion activity coefficients, $f_{ion,electrol}$, as

$$f^{\nu}_{\pm,electrol} = f^{\nu_+}_{+,electrol}f^{\nu_-}_{-,electrol} \tag{7a}$$

therefore,

$$f^2_{\pm,electrol} = f_{+,electrol}f_{-,electrol} \tag{7b}$$

Application of the information in Eqs. (6) and (7b) to Eq. (5) leads to

$$\left|\Delta\Phi_{electrol}\right| = \frac{i_{geom}l_{electrol}}{k_{electrol-eff}} + \left(\frac{RT}{F}\right)\left[2t_{Li^+}\ln\left(\frac{f^r_{\pm,electrol}c^r_{electrol}}{f^l_{\pm,electrol}c^l_{electrol}}\right) - \ln\left(\frac{f^r_{PF_6^-,electrol}c^r_{electrol}}{f^l_{PF_6^-,electrol}c^l_{electrol}}\right)\right] \tag{8}$$

The first term on the right-hand side of Eq. (8) is indicative of the ohmic voltage loss associated with the Li^+ ion movement in the electrolyte solution and the second term shows the effect of variation in the electrolyte salt concentration from the left to right side of the electrolyte solution region. For the situation of a dilute electrolyte solution or in the absence of reliable data on $f_{\pm,electrol}$ and $f_{PF_6^-,electrol}$, one may set $\frac{f^r_{\pm,electrol}}{f^l_{\pm,electrol}} \simeq 1$, and $\frac{f^r_{PF_6^-,electrol}}{f^l_{\pm,electrol}} \simeq 1$ in Eq. (8). This action leads to

$$\left|\Delta\Phi_{electrol}\right| = \frac{i_{geom} l_{electrol}}{k_{electrol-eff}} + \left(\frac{RT}{F}\right)\left(2t_{Li^+} - 1\right)\ln\left(\frac{c^r_{electrol}}{c^l_{electrol}}\right) \tag{9}$$

To evaluate $\left(\frac{c^r_{electrol}}{c^l_{electrol}}\right)$ in an electrolyte solution region, the following theoretical development is provided. The ion flux in the electrolyte solution region for an ion, $i = Li^+, PF_6^-$, at a z location is given by

$$N_{i,z} = \left(\frac{D_{i,electrol,eff}}{RT}\right) c_{i,electrol} \left(\frac{-d\mu_i^{echem}}{dz}\right) \tag{10}$$

where $D_{i,electrol,eff}$ is the effective ion mass diffusivity of an ion, i, through the electrolyte held in a cell layer and μ_i^{echem} is the electrochemical potential of an ion i in the electrolyte solution. The $D_{i,electrol,eff}$ is given by

$$D_{i,electrol,eff} = \frac{D_i \varepsilon_{electrol}}{\tau_{electrol}} = D_i \varepsilon^{1.5}_{electrol} \tag{11}$$

where D_i is the mass diffusivity of an ion i in the electrolyte solution, $\varepsilon_{electrol}$ the pore volume fraction of a cell layer containing the organic electrolyte solution, and $\tau_{electrol}$ the tortuosity factor of the cell layer containing the electrolytic solution. The ionic electrochemical potential, μ_i^{echem}, at a z location is usually expressed as the sum of chemical and electric potentials as follows:

$$\mu_i^{echem} = \mu_i^{chem} + z_i F \Phi \tag{12}$$

where the chemical potential of an ion, μ_i^{chem}, is given by

$$\mu_i^{chem} = \mu_i^\circ + RT\ln a_i = \mu_i^\circ + RT\ln\left(\frac{f_{i,electrol} c_{i,electrol}}{c_i^\circ}\right) \tag{13}$$

where μ_i° is the standard-state chemical potential of an ion species i at temperature $T\,[K]$. By the use of Eqs. (12) and (13) for the constant temperature condition, one obtains:

$$\frac{d\mu_i^{echem}}{dz} = \frac{d\mu_i^{chem}}{dz} + z_i F \frac{d\Phi}{dz} \quad (14)$$

$$= RT \frac{d}{dz}\left(\ln\left(\frac{f_{i,electrol} c_{i,electrol}}{c^\circ} \right) \right) + z_i F \frac{d\Phi}{dz} \quad (14a)$$

Substituting for $\frac{d\mu_i^{echem}}{dz}$ from Eq. (14a) in Eq. (10), on obtains, in conjunction with Eq. (13),

$$N_{i,z} = -D_i \varepsilon_{electrol}^{1.5} c_{i,electrol} \frac{d}{dz}\left[\ln\left(\frac{f_{i,electrol} c_{i,electrol}}{c^\circ} \right) \right] - \frac{D_i \varepsilon_{electrol}^{1.5}}{RT} c_{i,electrol} z_i F \frac{d\Phi}{dz} \quad (15)$$

The lithium ions from the anode lithium sheet, during the cell discharge, enter the organic electrolyte solution after migrating through the anode protective layer. Then, these lithium ions continue their journey through the electrolyte solution to the catalytic active sites in the porous, carbon cathode layer for the completion of the oxygen reduction process. The PF_6^- anion is neither produced nor consumed: only the Li^+ cation is the net charge carrier. Also, the local electrolyte solution charge neutrality is maintained. Equation (15) is applied to Li^+ and PF_6^- ions to obtain

$$N_{Li^+,z} = -D_{Li^+} \varepsilon_{electrol}^{1.5} c_{Li^+,electrol} \frac{d}{dz}\left[\ln\left(\frac{f_{Li^+,electrol} c_{Li^+,electrol}}{c^\circ} \right) \right] - \frac{\varepsilon_{electrol}^{1.5} D_{Li^+} c_{Li^+,electrol}}{RT} z_{Li^+} F \frac{d\Phi}{dz} \quad (16a)$$

$$0 = -D_{PF_6^-} \varepsilon_{electrol}^{1.5} c_{PF_6^-,electrol} \frac{d}{dz}\left[\ln\left(\frac{f_{PF_6^-,electrol} c_{PF_6^-,electrol}}{c^\circ} \right) \right] - \frac{\varepsilon_{electrol}^{1.5} D_{PF_6^-} c_{PF_6^-,electrol}}{RT} z_{PF_6^-} F \frac{d\Phi}{dz} \quad (16b)$$

Because $z_{Li^+} = +1$, $z_{PF_6^-} = -1$, and $c_{Li^+,electrol} = c_{PF_6^-,electrol} = c_{electrol}$, Eqs. (16a) and (16b) are reduced to

$$\frac{N_{Li^+,z}}{D_{Li^+}} = -\varepsilon_{electrol}^{1.5} c_{electrol} \frac{d}{dz}\left[\ln\left(\frac{f_{Li^+,electrol} c_{electrol}}{c^\circ}\right)\right] - \frac{\varepsilon_{electrol}^{1.5} c_{electrol}}{RT} F \frac{d\Phi}{dz} \tag{17a}$$

$$0 = -\varepsilon_{electrol}^{1.5} c_{electrol} \frac{d}{dz}\left[\ln\left(\frac{f_{PF_6^-,electrol} c_{electrol}}{c^\circ}\right)\right] + \frac{\varepsilon_{electrol}^{1.5} c_{electrol}}{RT} F \frac{d\Phi}{dz} \tag{17b}$$

Addition of Eqs. (17a) and (17b) leads to

$$\frac{N_{Li^+,z}}{D_{Li^+}} = -\varepsilon_{electrol}^{1.5} c_{electrol} \frac{d}{dz}\left[\ln\left(\left(\frac{c_{electrol}}{c^\circ}\right)^2 f_{Li^+,electrol} f_{PF_6^-,electrol}\right)\right] \tag{18a}$$

$$= -\varepsilon_{electrol}^{1.5} c_{electrol} \frac{d}{dz}\left[\ln\left(\frac{c_{electrol}}{c^\circ} f_{\pm,electrol}\right)^2\right] \tag{18b}$$

$$= -2\varepsilon_{electrol}^{1.5} c_{electrol} \frac{d}{dz}\left[\ln\left(\frac{c_{electrol}}{c^\circ} f_{\pm,electrol}\right)\right] \tag{18c}$$

$$= -2\varepsilon_{electrol}^{1.5} c_{electrol} \frac{d}{dz}\left[\ln c_{electrol} + \ln f_{\pm,electrol} - \ln c^\circ\right] \tag{18d}$$

For the case of dilute electrolyte solution or for the assumption of $\frac{d \ln f_{\pm,electrol}}{dz} \simeq 0$, Eq. (18d) simplifies to

$$\frac{N_{Li^+,z}}{D_{Li^+}} = -2\varepsilon_{electrol}^{1.5} c_{electrol} \frac{d \ln c_{electrol}}{dz} \tag{19a}$$

$$\frac{N_{Li^+,z}}{D_{Li^+}} = -2\varepsilon_{electrol}^{1.5} \frac{dc_{electrol}}{dz} \tag{19b}$$

The local geometric current density in a cell electrolyte containing layer is now given as

$$i_{geom} = N_{Li^+,z}F = -2FD_{Li^+}\varepsilon_{electrol}^{1.5}\frac{dc_{electrol}}{dz} \quad (20)$$

Under the steady-state condition in an electrolyte solution layer, with constant i_{geom} with respect to z, Eq. (20) is integrated with respect to z from $z = \frac{l_{electrol}}{2}$ to $z = z$ in the electrolyte solution layer.

$$\int_{c_{electrol}=c_{electrol,ave}}^{c_{electrol}=c_{electrol}} dc_{electrol} = \left(\frac{-i_{geom}}{2FD_{Li^+}\varepsilon_{electrol}^{1.5}}\right)\int_{z=\frac{l_{electrol}}{2}}^{z=z} dz \quad (21)$$

Equation (21) leads to:

$$c_{electrol} = c_{electrol,ave} - \frac{i_{geom}}{2FD_{Li^+}\varepsilon_{electrol}^{1.5}}\left(z - \frac{l_{electrol}}{2}\right) \quad (22)$$

which is valid for z in the electrolyte solution layer where i_{geom} is constant with respect to z. Now, $c_{electrol}^{r}$ is evaluated by setting $z = l_{electrol}$ in Eq. (22) to obtain

$$c_{electrol}^{r} = c_{electrol,ave} - \frac{i_{geom}l_{electrol}}{4FD_{Li^+}\varepsilon_{electrol}^{1.5}} \quad (23a)$$

Similarly, $c_{electrol}^{l}$ is obtained by setting $z = 0$ in Eq. (22) to yield

$$c_{electrol}^{l} = c_{electrol,ave} + \frac{i_{geom}l_{electrol}}{4FD_{Li^+}\varepsilon_{electrol}^{1.5}} \quad (23b)$$

Using Eqs. (23a) and (23b), one obtains:

$$\frac{c_{electrol}^{r}}{c_{electrol}^{l}} = \frac{1 - \dfrac{i_{geom}l_{electrol}}{4Fc_{electrol,ave}D_{Li^+}\varepsilon_{electrol}^{1.5}}}{1 + \dfrac{i_{geom}l_{electrol}}{4Fc_{electrol,ave}D_{Li^+}\varepsilon_{electrol}^{1.5}}} \quad (24)$$

This equation is truly valid in an electrolyte solution separator layer between the protective anode layer and the porous cathode layer. Substitution for $\left(\frac{c_{electrol}^{r}}{c_{electrol}^{l}}\right)$ from Eq. (24) into Eq. (9) leads to

$$\left|\Delta\Phi_{electrol}\right| = \frac{i_{geom} l_{electrol}}{k_{electrol-eff}} + \left(\frac{RT}{F}\right)\left(2t_{Li^+} - 1\right)\ln\left(\frac{1 - \dfrac{i_{geom} l_{electrol}}{4Fc_{electrol,ave} D_{Li^+} \varepsilon_{electrol}^{1.5}}}{1 + \dfrac{i_{geom} l_{electrol}}{4Fc_{electrol,ave} D_{Li^+} \varepsilon_{electrol}^{1.5}}}\right) \quad (25)$$

The validity of this equation to compute $\left|\Delta\Phi_{electrol}\right|$ requires that

$$\frac{i_{geom} l_{electrol}}{4Fc_{electrol,ave} D_{Li^+} \varepsilon_{electrol}^{1.5}} < 1, \text{ or } i_{geom} < \frac{4Fc_{electrol,ave} D_{Li^+} \varepsilon_{electrol}^{1.5}}{l_{electrol}} \quad (26)$$

Dependence of the lithium/air system performance on oxygen solubility, concentration, and its transport in organic electrolytes was experimentally shown [2, 3]. A diffusion-limited model for oxygen transport and its conversion in the porous cathode reaction layer is presented below.

The developed oxygen continuity equation, describing oxygen transport and its reduction in the porous carbon cathode, is given below.

$$\frac{\partial}{\partial t}(\varepsilon c_A) = \frac{\partial}{\partial z_1}\left(\frac{D_A \varepsilon}{\tau}\frac{\partial c_A}{\partial z_1}\right) - R_A \quad (27)$$

The oxygen conversion rate, R_A, at the active sites located at the pores surfaces per unit cathode volume is given by

$$R_A = \left(\frac{2\varepsilon}{\bar{r}_p}\right)(k_s c_A) = 2k_s\left(\frac{\varepsilon}{\bar{r}_p} c_A\right) \quad (28)$$

where k_s is the overall electrochemical oxygen reduction rate coefficient at a given cell operational temperature, and $\bar{r}_p$ is the mean pore radius of pores filled with the electrolyte solution. Note that here,

$$k_s = k_{s,e} \exp\left(\frac{\left|\eta^c\right| \alpha^c F}{RT}\right) = \left[k_{s,e,\circ} \exp\left(\frac{-E_\circ}{RT}\right)\right] \exp\left(\frac{\left|\eta^c\right| \alpha^c F}{RT}\right) \quad (29)$$

It is here assumed that there are n_p open-ended pores, each of mean pore radius, $\bar{r}_p$, corresponding to a unit geometric area perpendicular to the z_1 coordinate (see Figure 2 below). The initial pore volume in the cathode reaction layer just before the start of the oxygen reduction process per unit geometric area perpendicular to the z_1 coordinate is given by

$$\left(\pi \overline{r}_{p,\circ}^{2} \tau l_c\right) n_p = l_c \varepsilon_\circ \tag{30a}$$

The pore volume in the cathode reaction layer during the oxygen reduction process is given by

$$\left(\pi \overline{r}_{p}^{2} \tau l_c\right) n_p = l_c \varepsilon \tag{30b}$$

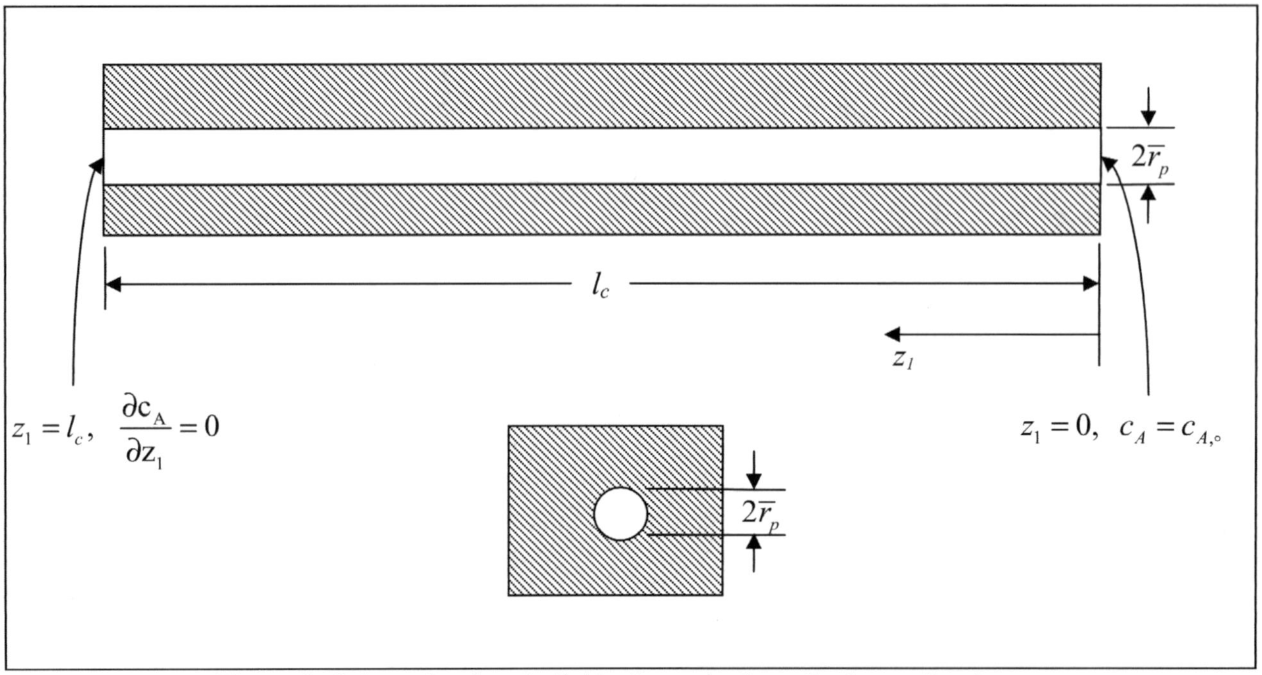

Figure 2. Schematic of an individual pore in the cathode reaction layer.

Division of Eq. (30b) by Eq. (30a) leads to

$$\varepsilon = \varepsilon_\circ \left(\frac{\overline{r}_p}{\overline{r}_{p,\circ}} \right)^2 \tag{31}$$

The following dimensionless quantities are now defined as:

$$r = \frac{\overline{r}_p}{\overline{r}_{p,\circ}}, \quad c_{A,d} = \frac{c_A}{c_{A,\circ}}, \quad \xi = \frac{z_1}{l_c} \tag{32a, b, c}$$

Substituting for R_A and ε from Eqs. (28) and (31), respectively, into Eq. (27); and expressing $\overline{r}_p$, c_A, and z_1 in terms of the dimensionless quantities, the following continuity equation for oxygen in the dimensionless form is obtained.

$$\frac{\partial \left(r^2 c_{A,d} \right)}{\partial t'} = \frac{\partial}{\partial \xi} \left(r^2 \frac{\partial c_{A,d}}{\partial \xi} \right) - \beta \left(r c_{A,d} \right) \tag{33}$$

where,

$$t' = \frac{t D_A}{\tau l_c^2} \tag{33a}$$

and

$$\beta = \frac{2k_s \tau l_c^2}{D_A \bar{r}_{p,\circ}} \tag{33b}$$

The rate of decrease of pore volume per unit total porous cathode volume due to the formation of solid $Li_2O_{2(s)}$ is given by

$$\frac{\partial \varepsilon}{\partial t} = -R_A \frac{M_{Li_2O_2}}{\rho_{Li_2O_{2(s)}}} \tag{34}$$

Substitution for R_A and ε from Eqs. (28) and (31), respectively, into Eq. (34); and further simplification leads to:

$$\frac{\partial \bar{r}_p}{\partial t} = -\frac{k_s c_A M_{Li_2O_2}}{\rho_{Li_2O_{2(s)}}} \tag{35}$$

Expressing $\bar{r}_p$ and t in the dimensionless form, Eq. (35) becomes

$$\frac{\partial r}{\partial t'} = -\gamma c_{A,d} \tag{36}$$

where,

$$\gamma = \left(\frac{k_s c_{A,\circ} M_{Li_2O_2}}{\rho_{Li_2O_{2(s)}}} \right) \Big/ \left(\frac{D_A \bar{r}_{p,\circ}}{\tau l_c^2} \right) \tag{36a}$$

To solve the coupled differential equations given by Eqs. (33) and (36) for $c_{A,d}$ and r, the following conditions were used.

Initial conditions, $t' \leq 0;\ \ 0 \leq \xi \leq 1$:

$$r = \frac{\bar{r}_p}{\bar{r}_{p,\circ}} = 1 \tag{37a}$$

$$c_{A,d} = \frac{c_A^\circ}{c_{A,\circ}} \tag{37b}$$

Boundary conditions, $t' > 0$:

at $\xi = 0$,

$$c_{A,d} = \frac{c_A}{c_{A,\circ}} = 1 \tag{38a}$$

at $\xi = 1$,

$$\frac{\partial c_{A,d}}{\partial \xi} = 0 \tag{38b}$$

Oxygen molar flux into the electrolyte at any time during the cell discharge period, per unit geometric area perpendicular to z_1, is given by

$$\dot{N}_{O_2}\Big|_{z_1=0} = \frac{D_A \varepsilon}{\tau} \left(-\frac{\partial c_A}{\partial z_1} \right)\Bigg|_{z_1=0} \tag{39a}$$

On substitution for ε from Eq. (31) into Eq. (39a) and the use of dimensionless quantities, one obtains

$$\dot{N}_{O_2}\Big|_{\xi=0} = \left(\frac{\varepsilon_\circ D_A c_{A,\circ}}{\tau l_c}\right)\left[r^2\left(-\frac{\partial c_{A,d}}{\partial \xi}\right)\right]\Bigg|_{\xi=0} \tag{39b}$$

The current generation rate per unit geometric area perpendicular to z_1 in a differential thickness of dz_1, via the occurrence of the electrochemical reaction: $2Li^+ + 2e^- + O_2 \rightleftharpoons Li_2O_{2(s)}$, is given by

$$di_{geom} = 2FR_A dz_1 \tag{40}$$

Using the expression for R_A in Eq. (28)

$$di_{geom} = \frac{4Fk_s \varepsilon c_A}{\bar{r}_p} dz_1 \tag{40a}$$

Substituting for $\left(\frac{\varepsilon}{\bar{r}_p}\right)$ from Eq. (31) into Eq. (40a) and using the dimensionless quantities given in Eqs. (32a, b, c), one obtains

$$di_{geom} = 4F\left(k_s c_{A,\circ}\right)\left(\frac{\varepsilon_\circ l_c}{\bar{r}_{p,\circ}}\right) r c_{A,d} d\xi \tag{41}$$

Total geometric current density, corresponding to the oxygen reduction reaction occurring in the cathode reaction layer is then given by

$$i_{geom} = \int_0^{i_{geom}} di_{geom} = \left[4F\left(k_s c_{A,\circ}\right)\left(\frac{\varepsilon_\circ l_c}{\bar{r}_{p,\circ}}\right)\right] \int_{\xi=0}^{\xi=1} \left(r c_{A,d}\right) d\xi \tag{42}$$

It should be noted that Eq. (42) gives oxygen reduction geometric current density at any time during the cell discharge. At the end of each time interval, the numerical solution of Eqs. (33) and (36) would provide data on the dimensionless oxygen concentration, $c_{A,d}$, and the dimensionless average pore radius, r, as a function of the dimensionless distance, ξ. Using this data in Eq. (42), one can carry out the integration of Eq. (42) either numerically or graphically to determine the average geometric current density at the end of each time interval. If i_{geom} is to be kept at a constant level with respect to time, Eq. (42) suggests that the product $\left[k_s \int_0^1 \left(r c_{A,d}\right) d\xi\right]$ must be kept constant. Because $\int_{\xi=0}^{\xi=1} \left(r c_{A,d}\right) d\xi$ is likely to decrease with time; therefore, k_s must be increased to a level so that $\left[k_s \int_0^1 \left(r c_{A,d}\right) d\xi\right] =$ constant with respect to time. From Eq. (42), in conjunction with Eq. (29), i_{geom} is expressed as:

$$i_{geom} = \left(4F\frac{\varepsilon_\circ l_c}{\bar{r}_{p,\circ}} c_{A,\circ}\right)\left(k_{s,e}\exp\left(\frac{\left|\eta^c\right| \alpha^c F}{RT}\right)\right) \int_{\xi=0}^{\xi=1} \left(r c_{A,d}\right) d\xi \tag{43a}$$

The geometric current density at a time, $t = t_1$ or $t' = t_1'$, is given by

$$i_{geom,t_1} = \left(\frac{4F\varepsilon_{\circ} l_c c_{A,\circ}}{\overline{r}_{p,\circ}} \right) \left(k_{s,e} \exp\left(\frac{\left|\eta^c\right|_{t_1} \alpha^c F}{RT} \right) \right) \left[\int_{\xi=0}^{\xi=1} \left(r c_{A,d} \right) d\xi \right]_{t_1} \tag{43b}$$

Division of Eq. (43a) by (43b) leads to:

$$\frac{i_{geom}}{i_{geom,t_1}} = \exp\left[\left(\left|\eta^c\right| - \left|\eta^c\right|_{t_1} \right) \frac{\alpha^c F}{RT} \right] \left[\frac{\int_0^1 \left(r c_{A,d} \right) d\xi}{\left\{ \int_0^1 \left(r c_{A,d} \right) d\xi \right\}_{t_1}} \right] \tag{44}$$

In order that the geometric current density can be kept constant with respect to time, i.e., $\left(\frac{i_{geom}}{i_{geom,t_1}} \right) = 1$, it requires that

$$\exp\left[-\frac{\left(\left|\eta^c\right| - \left|\eta^c\right|_{t_1} \right) \alpha^c F}{RT} \right] = \frac{\int_0^1 \left(r c_{A,d} \right) d\xi}{\left\{ \int_0^1 \left(r c_{A,d} \right) d\xi \right\}_{t_1}} \tag{45}$$

Equation (45) suggests that if the average geometric current density at any time, t, is to be equal to that at a time, $t = t_1$, where the oxygen reduction overvoltage is $\left|\eta^c\right|_{t_1}$; the overvoltage $\left|\eta^c\right|$ must be adjusted and equal to that given by Eq. (45). Because the numerator of the right-hand side of Eq. (45) will decrease as the cell discharge time increases, therefore, the cathodic reaction overvoltage, $\left|\eta^c\right|$, should increase with time to maintain the geometric current density at a desired constant level.

If $k_{s,e}$ and α^c which are needed to compute i_{geom} from Eq. (43a) are not available, they must be determined experimentally using a very thin porous carbon cathode under electrochemical kinetic control conditions. In fact, the reaction-order with respect to the dissolved oxygen concentration in the electrolyte solution can also be determined. The geometric current density in a thin porous carbon cathode of thickness $l_{c,thin}$, following Eq. (41), under the electrochemical reaction control conditions, is expressed as:

$$i_{geom} = 4F \left(\frac{\varepsilon_{\circ}}{\overline{r}_{p,\circ}} \right) \left(r l_{c,thin} \right) k_s c_{A,\circ}^m \tag{46}$$

$$= \left(4F \frac{\varepsilon_\circ}{\bar{r}_{p,\circ}} l_{c,thin} \right) r k_{s,e} \exp\left(\frac{|\eta^c| \alpha^c F}{RT} \right) c_{A,\circ}^m \tag{46a}$$

Therefore,

$$\frac{i_{geom}}{r} = \left(4F \frac{\varepsilon_\circ}{\bar{r}_{p,\circ}} l_{c,thin} \right) k_{s,e} \exp\left(\frac{|\eta^c| \alpha^c F}{RT} \right) c_{A,\circ}^m \tag{46b}$$

On taking the logarithm of both sides of Eq. (46b) and subsequent simplification leads to:

$$\log_{10}\left(\frac{i_{geom}}{r} \right) = \log_{10}\left(\frac{4F\varepsilon_\circ}{r_{p,\circ}} l_{c,thin} \right) + \log_{10} k_{s,e} + \left(\frac{\alpha^c F}{2.303RT} \right) |\eta^c| + m \log_{10} c_{A,\circ} \tag{47}$$

This equation suggests that i_{geom} and $r = \frac{\bar{r}_p}{\bar{r}_{p,\circ}}$ should be measured as a function of $|\eta^c|$ at a constant temperature, T, and $c_{A,\circ}$ to determine the charge transfer parameter α^c. For this measurement, the oxygen concentration should be high so that there is negligible resistance to oxygen mass transfer to the reaction active sites in the cathode pores. Then, $\log_{10}\left(\frac{i_{geom}}{r} \right)$ plotted versus $|\eta^c|$ should be a straight line with a slope equal to $\frac{\alpha^c F}{2.303RT}$. Thus, from this slope one can determine α^c. With the value for α^c known, one can determine the reaction order m with respect to dissolved oxygen concentration in the electrolyte by changing $c_{A,\circ}$ and measuring i_{geom}, r, and $|\eta^c|$ as a function of $c_{A,\circ}$ at constant T. Note that if the experiments are performed at the initial state conditions, $r \cong 1$ for a set of values of $\left(\frac{i_{geom}}{r} \right)$, $|\eta^c|$, and $c_{A,\circ}$, one can find m and also $k_{s,e}$ at a temperature. Using at least three sets of values of $\left(\frac{i_{geom}}{r} \right)$, $|\eta^c|$, and $c_{A,\circ}$ at three temperatures, one can calculate at least three values of $k_{s,e}$ as a function of a temperature. The following procedure can be carried out to determine the pre-exponential factor and reaction activation energy, $E_\circ$.

$$k_{s,e} = k_{s,e,\circ} \exp\left(-\frac{E_\circ}{RT} \right) \tag{48}$$

Taking $\log_{10}$ of both sides of Eq. (48) leads to:

$$\log_{10} k_{s,e} = \log_{10} k_{s,e,\circ} + \left(-\frac{E_\circ}{2.303R}\right)\left(\frac{1}{T}\right) \tag{49}$$

Equation (49) suggests that the plot of $\log_{10} k_{s,e}$ versus $\frac{1}{T}$ should be a straight line with the slope equal to $\left(-\frac{E_\circ}{2.303R}\right)$ and intercept equal to $\log_{10} k_{s,e,\circ}$. Hence, $k_{s,e,\circ}$ and $E_\circ$ can be determined for their application in Eq. (48) to compute $k_{s,e}$. In turn, $k_{s,e}$ can be used to calculate k_s from Eq. (29) for its application in Eqs. (33) and (36).

3. Results and Discussion

Under numerically constant geometric current density and oxygen solubility in the organic electrolyte solution, Eqs. (33) and (34) were solved numerically using an explicit finite difference technique. Table 1 lists the parameters used in the simulation.

Table 1. Parameter Values used in Simulation

Parameter	Value(s)
l_c	0.001, 0.010, and 0.070 cm
$c_{A,\circ}$	6.44692E-07 g-mol/cc at 0.2 atm O_2 1.934076E-06 g-mol/cc at 0.6 atm O_2 3.22346E-06 g-mol/cc at 1 atm O_2 (all for 1M $LiPF_6$ in PC/DME)
$\varepsilon_\circ$	0.73 [3]
D_A	7.00E-6 cm^2/sec [2]
$\bar{r}_{p,\circ}$	2.00E-7 cm (estimate)
$MW_{Li_2O_2}$	45.8768 g/g-mol
$\rho_{Li_2O_2}$	2.3 g/cc [4]
τ	1.17 (calculated from $\tau = \varepsilon_\circ^{-0.5}$)
i_{geom}	5, 1, 0.5, 0.1, 0.05 mA/cm^2
T	298.15 K
α	0.0722[2] for 1M $LiPF_6$ in PC/DME
α^c	1

Fig. 3 shows the dimensionless pore radius as a function of percent of total pore length and dimensionless time for a geometric current density of 0.5 mA/cm^2, a cathode thickness of 0.070 cm and an oxygen partial pressure of 1.0 atm. The highest dimensionless time represents the total discharge time for the cell. Fig. 4 has the same parameters as Fig. 3 except the geometric current density is 0.10 mA/cm^2. Note that at the lower current density, as shown by Fig. 4 in comparison to Fig. 3, a higher utilization of the electrode is realized as shown by the lower steepness of the dimensionless pore radius versus the dimensionless pore axial distance profile. Also note that the pore radius gradient in Fig. 4 is less than that for Fig. 3 due to the lower current density.

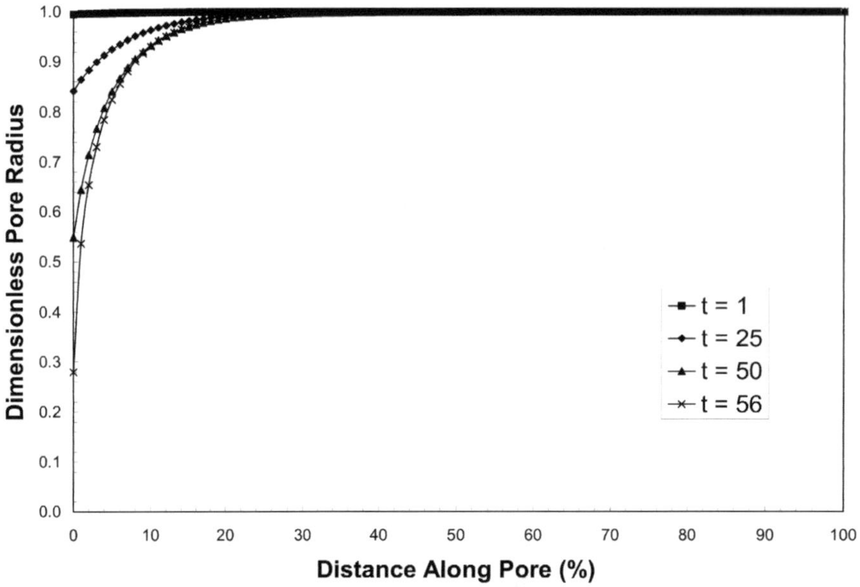

Figure 3. Dimensionless pore radius versus pore distance for i_{geom} = 0.5 mA/cm^2, l_c = 0.07 cm and O_2 partial pressure = 1.0 atmosphere.

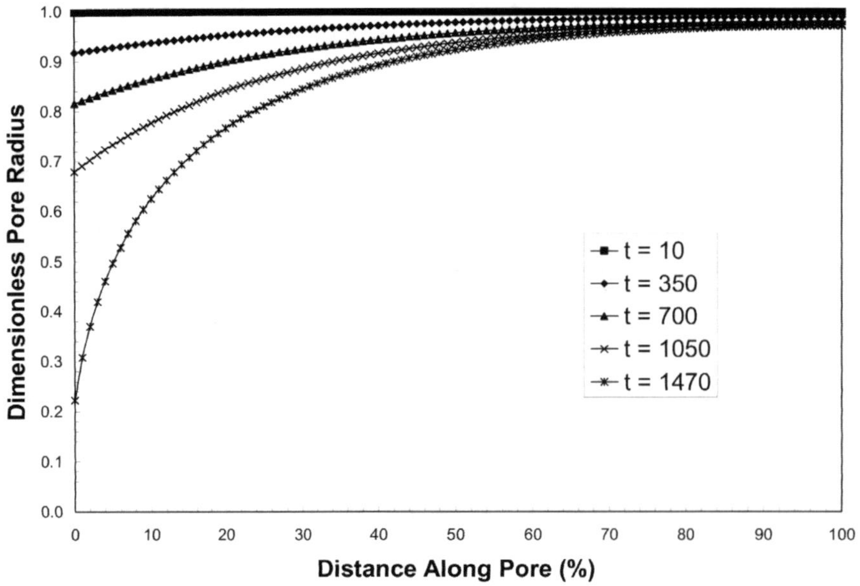

Figure 4. Dimensionless pore radius versus pore distance for i_{geom} = 0.1 mA/cm^2, l_c = 0.07 cm and O_2 partial pressure = 1.0 atmosphere.

With respect to Figs. 3 and 4, Figs. 5 and 6 show the corresponding dimensionless oxygen concentration profiles in the pores. Note that the concentration gradient in Fig. 5 is higher than Fig. 6 due to the higher current density. Also, the oxygen concentration limiting the discharge time is clearly evident in Figs. 5 and 6.

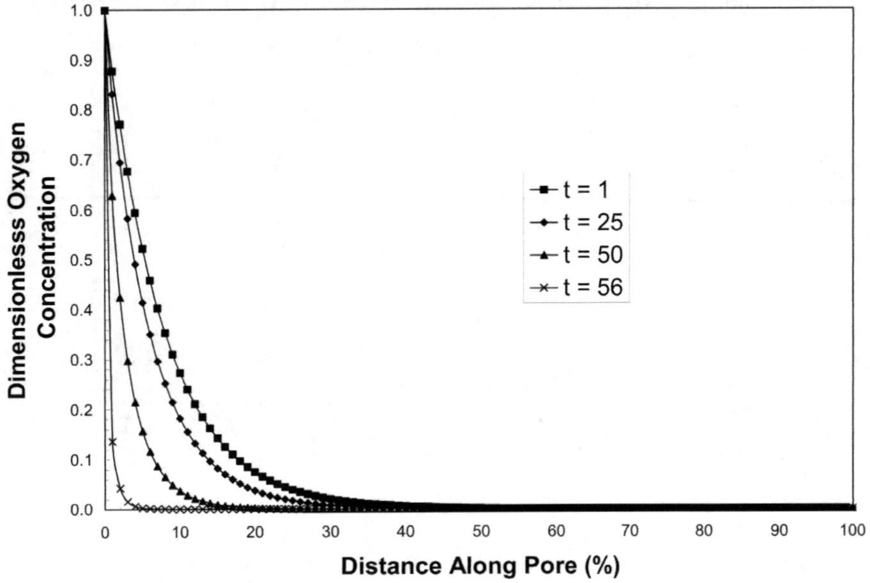

Figure 5. Dimensionless oxygen concentration versus pore distance for i_{geom} = 0.5 mA/cm^2, l_c = 0.07 cm and O_2 partial pressure = 1.0 atmosphere.

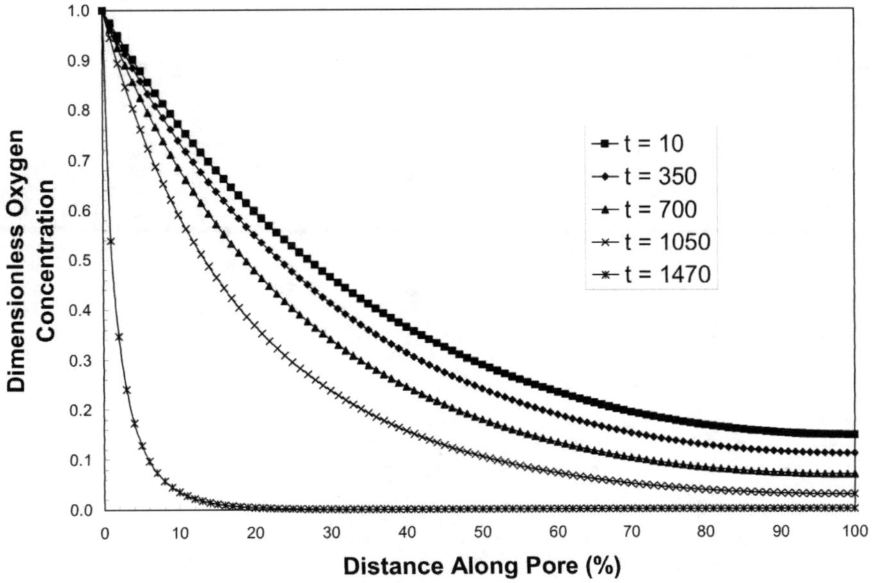

Figure 6. Dimensionless oxygen concentration versus pore distance for i_{geom} = 0.1 mA/cm^2, l_c = 0.07 cm and O_2 partial pressure = 1.0 atmosphere.

To determine the specific capacities (mAh/g of carbon) of the modeled cell, as a first approximation it was assumed that the solid volume fraction of the cathode, 0.27, was initially composed of only the carbon material with its density in the cathode being 2.2 g/cm^3. Fig. 7 shows the cell specific capacity as a function of oxygen partial pressure and geometric current densities. This curve is very similar to measured experimental values (see Fig. 3 in [2]).

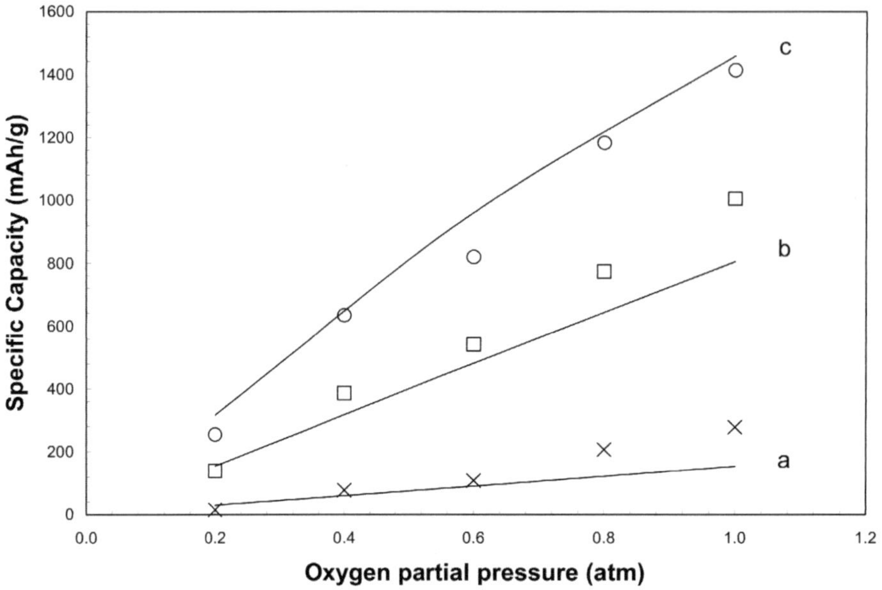

Figure 7. Lines are the calculated specific capacities (mAh/gram of carbon) for $l_c = 0.07cm$ as a function of oxygen partial pressure and current densities of (a) 0.5, (b) 0.1, and (c) 0.05 mA/cm^2. Symbols are the experimental values from Fig. 3 in [2].

Fig. 8 shows the specific capacity as a function of geometric current density and cathode thickness for this pore flooded-electrolyte design. Clearly, as the cathode thickness is reduced, the cell performance as measured in terms of specific capacity for a given current density is greatly enhanced.

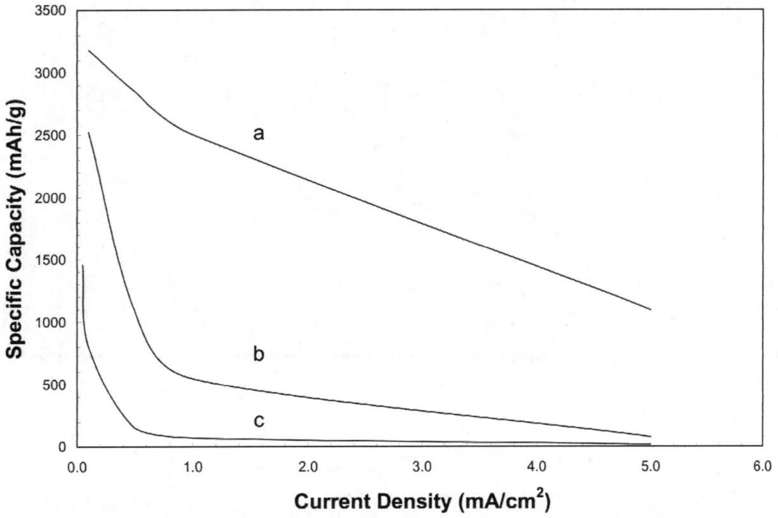

Figure 8. Specific capacity (mAh/g of carbon) as a function of current density for an oxygen partial pressure of 1.0 atm and l_c's of (a) 0.001, (b) 0.01, and (c) 0.07 cm.

It has clearly been shown above by simulation that Read's [2, 3] lithium/oxygen cells were flooded with electrolyte. The simulations represented by Fig. 7 are closely correlated with experimental results under the same conditions. Fig. 8 shows that thick cathode cell designs are not optimal and that a trade-off between starved and flooded electrolyte cell designs also needs to be performed.

4. Conclusions

A diffusion-limited model for lithium/air battery with an organic electrolyte was developed and the results from the model compare favorably with experimental results. The model shows, for a given cathode pore radius, the effect of oxygen partial pressure, current density, and cathode thickness on the specific capacity of the battery.

4. References

[1] J.S. Newman, Electrochemical Systems, Prentice Hall, Englewood Cliffs, N.J., 1991, 254 pp.

[2] J. Read, K. Mutolo, M. Ervin, W. Behl, J. Wolfenstine, A. Driedger, D. Foster, Oxygen Transport Properties of Organic Electrolytes and Performance of Lithium/Oxygen Battery, J Electrochem Soc 150(2003).

[3] J. Read, Characterization of the Lithium/Oxygen Organic Electrolyte Battery, J Electrochem Soc 149(2002) A1190-A1195.

[4] S.S. Sandhu, G.W. Brutchen, J.P. Fellner, Lithium-Air Battery System: Preliminary Analysis, 13th International Conference on Electrical and Electronic Products, Product Safety Corporation, Sissonville, West Virginia, 2006.

The Effect of Heat and Thermal Storage Capacities of Photovoltaic Duct Wall on Co-generation of Electric and Thermal Power

Himanshu Dehra, American Institute-Industry Forum on Energy, Lachine, Québec, Canada

Introduction

The generation of dissipated heat from a photovoltaic (PV) module is due to emitted short wave electro-magnetic radiation from its surface. Heat dissipation losses in PV module (~ 85%) result from several factors. Solar intensity is usually lower than the reference values of 1000 Wm^{-2}, as atmospheric absorption of the solar radiation spectrum reduces the available solar radiation below this value (Dehra, 2004). Electrical conversion efficiencies are also reduced because incident solar radiation is usually not perpendicular to the plane of the photovoltaic modules (Dehra, 2004). Moreover, solar cell temperatures are often high than standard test conditions, so that the diode current that opposes the light generating current is higher than in test conditions, reducing the output current. Electrical conversion efficiency decreases approximately by 1 % with every 25°C increase in surface temperature of crystalline silicon based solar cells (Dehra, 2004). In an effort to enhance overall efficiency of PV module power generation system, a novel solar energy utilization technique for co-generation of electric and thermal power is analyzed with a photovoltaic duct wall system. A full scale experimental facility for a photovoltaic duct wall was installed at Concordia University, Montréal, Canada (Dehra, 2004). The photovoltaic duct wall was comprised of a pair of glass coated PV modules, ventilated air passage and polystyrene filled plywood board. In this case duct wall with air ventilation acts as cooling channel for PV modules by reducing surface temperature of solar cells in PV modules and slightly increases its efficiency for electric power generation. With air as fluid medium, assessment of the potential use of photovoltaic duct wall to be used as a source of co-generation of electric and thermal power can be performed by thermal analysis of material properties of photovoltaic duct wall system (PV module, air and plywood board). The thermal analysis of photovoltaic duct wall will ascertain the amount of generated thermal power to be used as pre-conditioned fresh air along with conditioning from heating, ventilating and air-conditioning (HVAC) system.

The thermal analysis of a photovoltaic duct wall has been performed through experimental and numerical investigations. The measurement data collected from the experimental setup was for solar intensities, currents, voltages, air velocities and temperatures of air and composite surfaces. The measured temperatures were obtained as a function of height of photovoltaic duct wall. The heat transfer rate from a photovoltaic duct wall is a measure of heat storage and thermal storage capacities of its various components. The steady state heat transfer rate has been predicted by performing two dimensional energy and mass balances on discretised section of photovoltaic duct wall, to get solutions of one dimensional heat conduction and heat transport equations. The assumptions of steady state heat transfer and lumped heat capacity are validated by comparing heat losses along all major dimensions. The non-consideration of transient analysis has been justified by comparing thermal losses along all major dimensions.

Experimental Setup

The photovoltaic duct wall was installed on south facing façade of prefabricated outdoor room. The outdoor room was setup at Concordia University, Montréal, Québec, Canada for conducting practical investigations (Dehra, 2004). The photovoltaic duct wall was vertically inclined at 10° East of South on the horizontal plane. The test section of photovoltaic duct wall was assembled in components with two commercially available PV modules, air passage with air-gap width of 90 mm, plywood board filled with polystyrene as insulation panel, side walls made up of Plexiglas and all parts connected with wooden frames. The photovoltaic duct wall section was constructed with two glass coated PV modules each of dimensions: (989 mm X 453 mm). The PV modules were having glass coating of 3 mm attached on their exterior and interior sides. The plywood board was assembled with 7 mm thick plywood board enclosure filled with 26 mm polystyrene. The overall thickness of plywood board with polystyrene was 40 mm. The exterior dampers were made of wood covered with an aluminium sheet. The heating, ventilating and air-conditioning (HVAC) requirements were met in the outdoor room by a baseboard heater, an induced-draft type exhaust fan and a split window air conditioner (Dehra, 2004). The heating was supplemented by conditioning from the fresh air entering from the inlet damper through photovoltaic duct wall. However, during the mild season of autumn for the duration of conducting experimental runs, neither baseboard heater was used nor air-conditioning unit was used for auxiliary heating or cooling inside the pre-fabricated outdoor room.

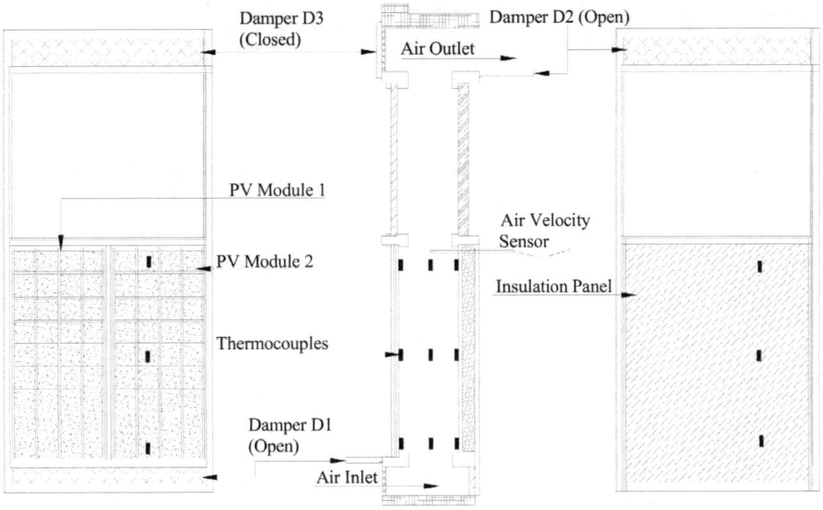

Fig. 1 Schematic of the Experimental Setup

The pair of PV modules used for conducting experimental investigations was connected in series for generation of electric power with a rheostat of maximum varying resistance up to 50 Ω. T-type thermocouples were used for obtaining thermal measurements from the test section of photovoltaic module. As is illustrated in Fig. 1, three thermocouple sensors were placed at the top, middle and bottom locations in the PV module, air-passage and insulation panel of plywood board filled with polystyrene were used to measure local temperatures. Two thermocouples were used to measure the inside test room air temperature and ambient air temperature. The hybrid air ventilation created for the PV module test section was by natural wind, or through buoyancy effect in the absence of wind (Dehra, 2004). The fan pressure was used to achieve higher air velocities by operation of the exhaust fan fixed on opposite wall with respect to wall of the test section (Dehra, 2004). The slight negative pressure was induced for drawing low air velocities in absence of wind-induced pressure from the inlet damper into the test section through the test room (Dehra, 2004). Air velocity sensor was placed perpendicular to the walls of the PV module test section to record axial air velocities near its outlet. The thermocouple outputs, currents, voltages, solar irradiation and air velocity signals were connected to a data logger and a computer for data storage. The measurements collected from the sensors were recorded as a function of air velocities or mass flow rate from the test section with use of fan pressure. The experimental data from the data acquisition system was collected and stored every two minutes in the computer (Dehra, 2004).

Numerical Investigations

Assumptions

The assumptions applicable to the simulation model are (Dehra, 2004): (i) quasi steady state heat transfer analysis has been performed for a photovoltaic duct wall assuming parallel plate configuration; (ii) uniform average air velocity distribution; (iii) temperature variation only in y-direction (vertical), being taken as lumped in other directions (x-axis and z-axis); (iv) air properties are evaluated at film temperature of 300 K; (v) negligible heat transfer from side walls/plywood board and room air zone; (vi) conduction (diffusion) equation for performing energy balance on air nodes is not taken into consideration; (vii) no infiltration or air leakage sources from the test section; and (viii) ambient air and room air temperatures are specified.

Governing equations

For photovoltaic module of the duct with a steady solar heat flux generation, Poisson's heat equation with boundary conditions is written as:

$$\frac{\partial^2}{\partial x^2} T_p + \frac{\partial^2}{\partial y^2} T_p + \frac{q_p}{k_p} = 0 \quad \text{in} \quad 0 < x < t_p \qquad 0 < y < H \tag{1}$$

$$\frac{\partial}{\partial x} T_p = \frac{\alpha\, S - h_a \left(T_p - T_a\right)}{k_p} \qquad \text{at} \quad x = 0 \tag{2}$$

$$\frac{\partial}{\partial x} T_p = \frac{h_p\left(T_p - T_a\right) + \sum_{i=1}^{n} hr_{p,i}\left(T_p - T_i\right)}{k_p} \qquad \text{at} \qquad x = t_p \tag{3}$$

For plywood board of duct with insulation, Laplace's heat equation with boundary conditions is written as:

$$\frac{\partial^2}{\partial x^2} T_i + \frac{\partial^2}{\partial y^2} T_i = 0 \qquad \text{in} \quad L + t_p < x < L + t_p + t_i \quad 0 < y < H \tag{4}$$

$$\frac{\partial}{\partial x} T_i = \frac{h_s \left(T_s - T_i\right)}{k_i} \qquad \text{at} \qquad x = L + t_p + t_i \tag{5}$$

$$\frac{\partial}{\partial x} T_i = \frac{h_i(T_i - T_a) + \sum_{p=1}^{n} hr_{p,i}(T_i - T_p)}{k_i} \qquad \text{at} \qquad x = L + t_p \tag{6}$$

Two Dimensional Energy and Mass Balance Equations

The energy balance between convective flux and enthalpy flux of air with its boundary condition is written as:

$$\theta\left(\frac{\partial}{\partial y} T_a\right) + T_a - \left(\frac{T_p + T_i}{2}\right) = 0 \qquad \text{in} \quad 0 < y < H \tag{7}$$

$$T_a(y) = T_a(k) \qquad \text{at} \quad y = \left(k + \frac{1}{2}\right)\cdot\frac{H}{n} \qquad k = 0..(n-1) \tag{8}$$

$$\theta = \frac{v \rho L c_p}{h_c W} \tag{9}$$

The conjugate heat exchange analysis is performed for a section of photovoltaic duct wall with one PV module and corresponding areas of air passage and insulating plywood board. The finite difference approximations are used to develop algebraic nodal equations for each node in the computational grid. The nodal equations are developed by performing an energy balance on surface and air nodes in the control volume i.e. control volume finite difference method (CVFDM) is used for representing algebraic nodal equations (Dehra, 2004). The general energy balance equation for 'p' surface nodes of PV module is written as (Dehra, 2004):

$$(T_o - T_a)\, U_{p,a} + q_p\, \Delta A_y + (T_p - T_{p+1})\, U_{p,p+1} + (T_p - T_{p-1})\, U_{p,p-1} + (T_p - T_a)\, U_{p,a} + \sum_i U_{p,i}(T_p - T_i) = 0 \tag{10}$$

Similarly, the energy balance equation for 'i' surface nodes of insulating plywood board is written as:

$$(T_i - T_s)\, U_{i,s} + (T_i - T_{i+1})\, U_{i,i+1} + (T_i - T_{i-1})\, U_{i,i-1} + (T_i - T_a)\, U_{i,a} + \sum_p U_{i,p}(T_i - T_p) = 0 \tag{11}$$

The energy balance equation for 'a' air nodes is written as:

$$U_{a,a-1}(T_a - T_{a-1}) + U_{a,i}(T_a - T_i) + U_{a,p}(T_a - T_p) = 0 \tag{12}$$

The solution of nodal equations of thermal network comprises formulation of conductance's matrix and heat source matrix (Dehra, 2004). Matrix equation has heat source vector terms for Q's and

U's are the conductance terms at various nodes. The general matrix form of nodal equation is written as (Dehra, 2004):

$$\sum_{n=1}^{N} \left(U_{m,n} \times \Delta T_{m,n}\right) + \sum_{n=1}^{N} Q_{m,n} = 0 \tag{13}$$

Results and Discussions

Temperature Solutions

The sample measurement results obtained from section of photovoltaic duct wall are presented in Table 2. The location of sensors as per Fig. 1 is presented in Table 3. The measurements that were obtained from experimental setup of photovoltaic duct wall are for cases of buoyancy-induced and fan-induced hybrid ventilation. The air temperatures in photovoltaic duct wall are developed as a function of magnitude of the ambient air temperature. The maximum convection heat transfer was observed between the ambient cold air entering into the inlet damper and the air passage of the test section. The dissipation of heat at the inlet opening has resulted in considerable increase in temperature of the air. For the general case fan-induced hybrid ventilation, the temperature difference between top and bottom node for PV module was around 2°C, for insulating plywood board between 8-10°C and temperature difference between ambient air and air leaving the outlet of photovoltaic duct wall varies between 5-9°C (Dehra, 2004). The insulating plywood board was heated by long wave radiation heat exchange from PV module and its temperature at top node was ~ 15°C higher than ambient air and surface temperature difference between top and bottom nodes of the insulation panel was ~ 10°C (Dehra, 2004). The temperature differences were higher for the case of buoyancy-induced hybrid ventilation. Thermal power is easily calculated for various components of photovoltaic duct wall from the temperature differences.

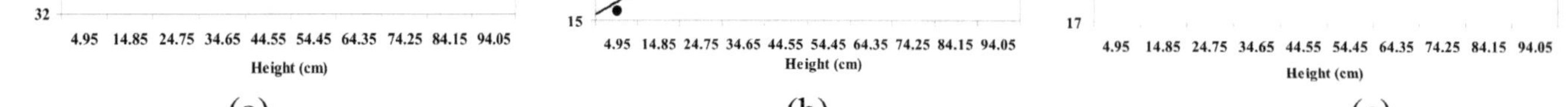

(a) (b) (c)

Fig. 2 Temperature plots with height of photovoltaic duct wall: (a) PV module; (b) Air; (c) Insulating plywood board

The results obtained from two dimensional numerical solutions for Run no. 1 of Table 1 are illustrated from Fig. 2(a) to Fig. 2(c). The temperature profiles of PV module, air and insulation panel are obtained with respect to the height of photovoltaic duct wall. The initial boundary values in the model are based on the measured values of quasi steady-state solar irradiation, ambient air temperature, and room air temperature and air velocities. The measured boundary values were then used to obtain two dimensional numerical solutions for one dimensional heat equations of the photovoltaic duct wall. The numerical results are obtained by assuming constant air and surface properties. The mean air velocity was assumed be constant in the test section for representing nodes of air with same thermal capacity conductance terms, mc_p also termed as thermal capacity rate C_p through out its height (Dehra, 2004). The amount of dissipated heat from the PV module taken up by air is dependent on mass flow rate passing over it, thermal capacity of air and temperature difference between air-column in the test section (Dehra, 2004). The dissipated heat from the face area (W X H) of PV module is ventilation heat rate is convection heat loss from the walls of PV module test section due to difference in temperature between the walls of test section and air flowing across cross-sectional area of test section across one PV module (Dehra, 2004).

Table 1 Sample Experimental Results

Run No.	S ($W m^{-2}$)	E_p (W)	T_o (° C)	T_s (° C)	V ($m s^{-1}$)	$T_p(b)$ (° C)	$T_p(m)$ (° C)	$T_p(t)$ (° C)	$T_b(b)$ (° C)	$T_b(m)$ (° C)	$T_b(t)$ (° C)	$T_a(a)$ (° C)	$T_a(m)$ (° C)	$T_a(t)$ (° C)
Fan-induced hybrid ventilation														
1	725.4	31.1	14.5	22.6	0.425	34.2	32.9	36.3	19.9	24.0	29.4	18.8	19.2	22.4
Buoyancy-induced hybrid ventilation														
2	697.5	28.8	13.3	24.9	0.17	39.9	45.0	46.8	28.4	35.0	38.3	21.7	28.3	29.8

Table 2 Location of Sensors in the Experimental Setup

Distance as per location of sensors (Fig. 2)	$T_p(b)$ (° C)	$T_p(m)$ (° C)	$T_p(t)$ (° C)	$T_b(b)$ (° C)	$T_b(m)$ (° C)	$T_b(t)$ (° C)	$T_a(b)$ (° C)	$T_a(m)$ (° C)	$T_a(t)$ (° C)	Air velocity sensor
y (cm)	15	55	94	15	55	94	15	55	94	99
z (cm)	60	60	60	60	60	60	60	60	60	60
x (mm)	6.2	6.2	6.2	96.2	96.2	96.2	51.2	51.2	51.2	51.2

Note: x is horizontal; y is vertical; z is adjacent 3rd axis of x-y plane of Fig. 1

Sensible Heat Storage Capacity

In Table 3, heat storage capacity has been calculated both for temperature differences in y and x-directions. Thermo-physical properties of various components of photovoltaic duct wall were evaluated at 300 K. Thermo-physical properties of plywood board with insulation were obtained from tests conducted with heat flow meter and related specifications from the manufacturer. The temperature differences along y-direction are obtained from the two dimensional numerical model. The temperature differences along x-direction are obtained by assuming same temperature difference per unit thickness of material along x and y-directions. The photovoltaic duct wall with wooden frame was composed of non-homogeneous materials having different densities, specific heats and thicknesses. The pair of PV module was having three layers of material viz., a flat sheet of solar cells, with glass face sheets on its exterior and interior sides. The surface temperature of PV module was assumed to be uniformly distributed in the three layers. The heat capacity of the wooden frame and sealing material was having negligible effect on the temperature of PV module, air or insulation panel because wood was used as construction material and moreover the magnitude of the heat capacity of wood framing material was not proportional to the face area of glass coated PV modules.

Table 3 Sensible Heat Storage Capacities

Component	ρ_n (kg m^{-3})	C_n (J Kg^{-1} K^{-1})	d_n (m) X 10^{-3}	$d_n\rho_nC_n$ (J m^{-2}K^{-1})	H_{pv-T} (J K^{-1})
Glass coating	3000	500	3	4500	4171.5
PV module	2330	677	0.2	315.48	292.45
Glass coating	3000	500	3	4500	4171.5
Sub-total	-	-	-	-	8635.5
Air	1.1174	1000	90	100.56	93.22
Plywood	550	1750	7	6737.5	6245.66
Polystyrene	1050	1200	26	32760	30368.5
Plywood	550	1750	7	6737.5	6245.66
Sub-total	-	-	-	-	42953.0
Total	-	-	-	-	51588.5

Note: Heat capacities are calculated for face area of PV module test section of 0.927 m^2.

Thermal Storage Capacity

Thermal storage capacity of photovoltaic duct wall has been obtained from individual thermal conductivities of glass coating, solar cells, air, plywood board and polystyrene. Time constant ($T=\rho_dC_{pd}d_d/h_d$) for each component of duct is calculated from individual heat capacities and film coefficients for various component of photovoltaic duct wall. Equivalent thermal conductivity of glass coated PV module is calculated to be 0.91 W m^{-1} K^{-1}. Temperature differences are obtained along y-direction i.e. along height of PV module test section (0.993 m) are obtained from Table 1 for Run No. 2 in the critical case of buoyancy-induced hybrid ventilation. Temperature differences along x-direction i.e. along thicknesses of each component of PV module test section are obtained proportionate to temperature differences along y-direction. Thermal storage capacity of PV module test section along its height is 15.9 KJ. Thermal storage capacity in x-direction is negligible in comparison with the thermal storage capacity in y-direction. Therefore temperature measurements were also felt necessary along the height of photovoltaic duct wall to consider pattern of heat flow and heat transport. The procedure for obtaining thermal storage capacities of components of photovoltaic duct wall is described in Table 4.

Table 4 Thermal Storage Capacities

Component	k_d (W m^{-1} K^{-1})	$d_n\rho_nC_n$ (J m^{-2}K^{-1})	H_d (Wm^{-2}K^{-1})	T (sec)	ΔT_V (K)	ΔT_H (K)	Q_V (KJ)	Q_H (J)
PV module	0.91	9315.48	10	932	6.9	0.04	5.8	0.2
Air	0.02624	100.56	10.0	10	8.1	0.75	0.0	0.0
Plywood	0.0835	6737.5	10.0	674	9.9	0.40	0.55	0.16
Polystyrene	0.02821	32760	1.0	32760	9.9	0.40	9.0	9.6
Plywood	0.0835	6737.5	10.0	674	9.9	0.40	0.55	0.16
Total	-	-	-	-	-	-	15.9	10.12

Traditional one dimensional (1-D) temperature solutions obtained from Eqs. (1) to (7) does not take into account heat flow conductance across all thicknesses towards x, y and z axes. The error induced by not considering conduction heat flow analysis along height of photovoltaic duct wall (y-direction) is found to be significant. In addition to this, traditional 1-D solutions do not take into effect integrated radiation heat exchange in between the composite nodes. The small increase or decrease in surface temperature causes exponential increase or decrease in temperature of air. From Eqs. (10) to (13), conjugate heat exchange paths are bi-directional in U-matrix to allow adjustment of any imposed errors due to calculation of constitutive relations (or boundary conditions) with use of temperatures obtained from the traditional 1-D temperature solutions.

Heat storage capacity of photovoltaic duct wall is considered as nil by assuming lumped temperature distribution. The error induced by neglecting heat storage capacity is small because of discretisation procedure. As the induced error due to non consideration of heat capacity is dependent on grid size, therefore with increased number of nodes in the grid, the induced error is negligible by not considering heat storage capacity.

The constitutive relations for conductance terms in Eqs (10)-(13) are calculated over discretised control areas in y-z plane for conductive heat flow, radiation and convective heat exchange factors. Heat transport conductance terms are calculated from mass flow rate crossing the control volume in x-z plane i.e. same throughout photovoltaic duct wall assuming no leakage or infiltration sources. The two dimensional numerical solutions have considered the effect of thermal storage by incorporating conduction heat flow factors in y-direction. The error induced by not considering heat storage capacity is negligible in comparison with the error induced by not considering thermal storage in the walls of photovoltaic duct. The glass coated photovoltaic modules have low specific heat in comparison with air and insulated plywood board filled with polystyrene. This results in small value of heat capacity for the semi-conductor material of photovoltaic modules in comparison with plywood board with polystyrene insulation. The plywood board and polystyrene have high specific heats and are widely used for building insulation. The low thermal conductivities of plywood board and polystyrene give poor thermal storage in comparison to glass coated photovoltaic modules. The counter-effects of good heat storage capacity and poor thermal storage in plywood board and poor heat storage capacity and good thermal storage in photovoltaic modules are balanced by considering the conduction heat flow along y-direction both for outer and inner wall of photovoltaic duct wall in obtaining two dimensional numerical solutions. Therefore, it is definitely not essential to obtain solutions of two-dimensional steady heat conduction equations (Eqs. 1 and 4) in comparison with the necessity of performing transient analysis for the case where inner wall of the duct is massive and has high heat storage capacity. Therefore, only effective boundary conditions (Eqs. 3 and 4) are utilized for obtaining two dimensional solutions of one dimensional heat conduction and heat transport equations.

Conclusions

A study has been conducted to consider the effect of heat storage capacity and thermal storage capacity of photovoltaic (PV) duct wall on co-generation of electric and thermal power. The duct wall was built with a pair of glass coated PV modules installed on a wooden frame of a duct wall, leaving air passage with the wall section of plywood board filled with polystyrene. The steady state solution of one dimensional heat conduction and heat transport equations were obtained by performing two dimensional energy and mass balances on section of PV duct wall. The heat and thermal storage capacities of PV duct wall were obtained for comparing the errors and validating the assumptions made in the proposed model. It was envisaged that assumptions were valid because of negligible computational errors due to steady state solution. The study has concluded that for PV duct wall, two dimensional transient solution of heat conduction equation is not essential for considering the effect of heat and thermal storage capacities. The model has predicted fairly well the temperatures along the height of PV duct wall in comparison to the measurements obtained from the test section.

Nomenclature

$\Delta T_{m,n}$	Temperature difference at node m,n (K)
A	Face area of PV module (m^2)
c_p	Specific heat of air at constant pressure (J kg^{-1} K^{-1})
E_p	Electric power generated
H	Height of photovoltaic duct wall (m)
h_a	Ambient heat transfer coefficient (W m^{-2} K^{-1})
h_i	Convective heat transfer coefficient for insulating plywood board (W m^{-2} K^{-1})
h_c	Average convective heat transfer coefficient (W m^{-2} K^{-1})
h_p	Convective heat transfer coefficient for PV module (W m^{-2} K^{-1})
hr	Linearised radiation heat transfer coefficient (W m^{-2} K^{-1})
h_s	Film coefficient for the insulating plywood board facing room air zone (W m^{-2} K^{-1})
L	Air gap width of duct (m)
m	Steady mass flow rate (kg s^{-1})
n	Number of discretised elements in y-ordinate
N	Total number of discretised elements in grid
$Q_{m,n}$	Heat source term at the node (m,n)
S	Incident solar intensity (W m^{-2})
$U_{p,i}$	Conductance between nodes p and i (direction p to i) (W K^{-1})
$U_{m,n}$	Conductance Term at node (m,n),
S_p	Absorbed solar radiation by PV module (Watts)
$T_{m,n}$	Temperature matrix for discretised nodes (m X n)
T_a	Temperature variable for the air (K)
T_i	Temperature variable for the back panel (K).
t_i	Thickness of Back panel (m)
T_o	Ambient air temperature (K)
T_p	Temperature variable for the PV module (K)
t_p	Thickness of PV module (m)
T_s	Room air temperature (K)
v	Air velocity (m s^{-1})
W	Width (m)
Subscripts	
p	PV module
i	Insulating plywood board
a	Air
m	Position of row in U-matrix
n	Position of column in U-matrix
o	Ambient air
s	Room

References

1. Dehra, H. (2004), "A Numerical and Experimental Study for Generation of Electric and Thermal Power with Photovoltaic Modules Embedded in Building Façade", submitted/unpublished Ph.D. thesis, Department of Building, Civil and Environmental Engineering, Concordia University, Montréal, Canada, August 2004.

Production-Data Based Energy Modeling and Support Tool for Optimization of the Energy Consumption in Chemical Multipurpose Batch Plants

Andrej Szijjarto[1], Ulrich Fischer[2], and Konrad Hungerbühler[2]. (1) Institute for Chemical- and Bioengineering, ETH Zurich, Hoenggerberg, HCI G138, Zürich, Switzerland, (2) ICB, ETH Zurich, Hoenggerberg, HCI G138, Zürich, Switzerland

With growing energy prices, the energy consumption becomes an important cost factor which in multipurpose batch plant can reach as high as 10 % of overall costs. The energy efficiency improvement which was driven by oil crisis in the 1970's is nowadays gradually shifted from the financial to environmental level (Kyoto protocol and engagements of the companies towards local authorities to reduce CO2 intensity of the production). Energy efficiency improvement of the chemical facilities requires support tools for decision makers in order to make quick and competent analysis of the current situation and indicate the highest energy saving potentials. A novel approach was developed using production data in electronic format as an input for the energy models allowing modeling of the energy consumption with minimal engineering effort while delivering highly detailed results. The energy utility consumption was corrected by the empirical loss term which characterizes the energy efficiency of the particular apparatus.

The set of production data obtained from the case study plant for a specified period representing particular product portfolio was preprocessed in order to get the durations and process parameters (temperature, pressure, mass levels) of the particular processes which are conducted in specific equipment units. The input data include the deviations in the process parameters and the durations of the operations for different batches and their influence on the overall utility consumption can be observed. The energy consumption of the chemical process contains two terms – theoretical consumption (calculated from the production data) and the loss term. The loss terms for particular apparatuses were determined based on the measurement of the energy utility consumption and a database with the loss terms for particular apparatus was created (can be used for energy consumption modeling in other chemical multipurpose batch plants without necessity of measuring energy consumption for individual apparatus).

The result of the model allows to analyze the energy utility consumption batchwise and to identify the batch with minimal utility consumption. The analysis of the results was carried out in order to identify the operations with highest energy saving potential and possible optimization of the energy consumption.

Phase Change Heat Transfer in Cylinders

Subrahmaniyam Susarla, CHEMICAL ENGINEERING, OSMANIA UNIVERSITY,UNIVERSITY COLLEGE OF TECHNOLOGY, -, -, HYDERABAD-500007, India

Phase change heat transfer is an important phenomena in energy storage systems and has also solar applications.In this work, freezing of phase change materials(PCM) in copper cylinders would be presented.The freezing process with super heat in the melt is significantly affected by free convection resulting in lower heat transfer rates compared to the absence of initial super heat.As the orientation of test cylinder influences the convective heat transfer coefficients,our experimental study included both horizotal and vertical orientations of test cylinders. Two PCMs namely n-octadecane(MP 28C) and paraffin wax(MP60C)were used in the study.Range of parameters were: L/d:1 to 5,initial superheat:2 to 25C,temperature drop across frozen layer:2 to 28C,volume of PCM:50cc to 1000cc. Cooling was affected by immersion in isothermal water bath(accuracy<1C).Experimental data of radial and axial temperature distributionns in PCM were obtained with pt-100 sensors and frozen mass vs time data were obtained by direct weighing method.A discussion of results of temperature distribution and freezing rate variation with time and comparison with mathematical models avalable in literature would be presented.

Centrifuge or Filter-- Sterol Removal from Biodiesel

Theodore Neuman, Food Tec, Westfalia Separator Inc, 100 Fairway Ct., Northvale, NJ 07647

Biodiesel produced from soybean and palm olein feedstocks has shown an impurity identified as a Sterol/ glycoside which causes plugging of engine filters, especially in colder climates. The paper will discuss the various centrifufge/clarifier and filtration techniques employed in European and North American production facilities to remove the impurity.

Catalytic Conversion of Sugars and Glycerin to Biochemicals

Aiguo Liu, Christopher Luckett, Shannon Carroll, Peter Hogue, and Wayne Turbeville. Sud-Chemie, Inc, 1600 W. Hill St., Louisville, KY 40232

Sugars from fermentation of cellulosic materials and crude glycerin from fast growing biodiesel industry are important sources of biochemicals in an effort to substitute petroleum products. Various processes such as oxidation, hydrogenation, dehydration, hydrogenolysis can be utilized in transforming these raw materials of renewable sources into high-valued chemicals. Highly selective and efficient catalysts play a critical role for all these processes. Materials that are able to withstand high temperature and pressure reactions in an aqueous phase are also an essential requirement.

This talk will present general aspects of Sud-Chemie's efforts in catalyst research and development for these purposes, with emphases on conversion of glycerin and sorbitol to 1,2-propanediol, and other emerging biochemicals.

GLYCEROL TO GASOLINE CONVERSION

Trung Hoang, Grant Ballantyne, Tanate Danuthai, Lance L. Lobban, Daniel E. Resasco, Richard G. Mallinson

School of Chemical, Biological, and Materials Engineering, University of Oklahoma, 100 E. Boyd St., Norman, OK, 73019

ABSTRACT

Conversion of renewable feedstocks to biofuels has gained attention for commercialization and further research. Glycerol is a byproduct in the production of biodiesel from transesterification of vegetable oil. Conversion of glycerol to fuel is of interest in terms of utilizing a large byproduct for production and upgrading of biofuels. Glycerol to gasoline conversion was performed using acid catalysts, including HZSM-5, HZSM-11, ZSM-22, HY, HMOR, and H-OMEGA. The results show that the pore geometry of the zeolite affects the product distribution obtained from the reaction. Three dimensional pore zeolites, for instance, HZSM-5, HZSM-11, and HY zeolites tend to produce aromatics. Although, the pressure was elevated to 300 psig, aromatics were still produced from the three dimensional zeolite catalysts. However, no aromatics were produced from 20%HMOR, 40%HMOR, and H-OMEGA, all with one dimensional pore structures. The products observed from the reaction included formaldehyde, acetaldehyde, acetol, etc. Cyclic ethers were produced from all catalysts tested when ethylene glycol was used as reactant feed.

INTRODUCTION

In a time of less available petroleum reserves and increased awareness of environmental impact, much attention has been drawn to renewable energy sources, especially biofuels. In the production of biodiesel, a large amount of glycerol is formed as a byproduct when the fatty acids are removed from the triglycerides via transesterification. Although, glycerol is currently utilized in the pharmaceutical and food industries, its volume will greatly exceed the market if it is produced in large quantities from biodiesel processes. This project thus has the object of the utilization of glycerol to upgrade to products that would be fungible with gasoline.

The existing process known as the Methanol to Gasoline (MTG) process for the conversion of methanol to hydrocarbons and water was announced by Mobil long ago. The hydrocarbons obtained in this reaction are alkanes in the gasoline boiling range (C_4-C_{10}) but with substantial amounts of aromatics. The zeolite A and Erionite were used as the catalysts. In 1979, Chang and coworkers reported the reaction pathway for the conversion of methanol over HZSM-5. The reaction schemes included three main sequential steps as follows; (1) the dehydration of methanol to dimethyl ether; (2) the dehydration of dimethyl ether to olefins; (3) the transformation of olefins to paraffins and aromatics [Chang et al., 1979].

Recently, Gayubo et al. reported the successful transformation of other alcohols and oxygenates, such as, propanol, butanol, acetone, and butanone to higher paraffins and olefins over HZSM-5 catalyst [Gayubo, 2004]. Their proposed reaction schemes involved the dehydration of oxygenates to corresponding olefins and paraffins as the first step, followed by oligomerization.

Based on previous technology and literature, it seems promising to investigate the conversion of glycerol, an oxygenate containing three hydroxyl groups, into gasoline range hydrocarbons or oxygenates. In this contribution, the products obtained from screening catalysts HZSM-5, HZSM-11, HZSM22, HY, HMOR, and H-OMEGA are reported. The effects of pressure and temperature were also investigated.

EXPERIMENTAL

HZSM-5 and H-OMEGA zeolites were supplied by SUD-CHEMIE. HY and Na-MOR zeolites were supplied by Zeolyst International. HZSM-11 was synthesized using the procedure indicated in a patent [Chu, 1973], with some modification. Na^+ ions in the Na-MOR zeolites were replaced with NH_4^+ ions by ion exchange with 20 wt% and 40% NH_4NO_3 solution at 80°C for 10 h to obtain 20%HMOR and 40%HMOR samples, respectively. ZSM-22 zeolite was synthesized by following the procedures indicated in the literature [Earnst et al., 1989]

Catalytic activity measurements were carried out in stainless steel reactors at 300 °C and 400°C under nitrogen atmosphere at W/F of 0.2 and 0.5 h. The reaction took place under either atmospheric pressure or 300 psig. Prior to the reaction, the catalyst was pretreated *in situ* under flowing nitrogen for 1 h. Glycerol or ethylene glycol were fed using a syringe pump or high pressure piston pump. The products were analyzed using an online FID gas chromatograph equipped with a capillary INNOWAX column. The products were identified by using a Shimadzu gas chromatograph equipped with mass spectrometer (GC/MS).

PRELIMINARY RESULTS

Products obtained from the three dimensional pore zeolites

Products from the conversion of glycerol and ethylene glycol over HZSM-5, HZSM-11, and HY catalysts are summarized in Table 1. It was found that HZSM-5 and HY produced carbonyl compounds, for instance, acetaldehyde, formaldehyde, acrolein, and acetol from the conversion of glycerol at 300 °C. However, only aromatics were found for glycerol conversion on HZSM-5 and HY at 400 °C under atmospheric pressure. The aromatics observed from the reaction included benzene, toluene, C_8 aromatics, and C_9 aromatics. In addition, aromatic compounds were still formed when the pressure was elevated up to 300 psig. This behavior was also observed on HY zeolite catalyst.

The conversion of ethylene glycol, containing two hydroxyl groups, was also investigated to understand the role of dehydration. It was found that cyclic ethers were formed from the conversion of ethylene glycol over HZSM-5 and HY at 300 °C under atmospheric pressure. Aromatics were observed only when the temperature was increased to 400 °C. Interestingly, the conversion of ethylene glycol on HZSM-11 yielded carbonyls and aromatics although the reaction temperature was at 300 °C, as shown in Table 1.

Table 1. Products observed from HZSM-5, HZSM-11, and HY zeolite catalysts at various conditions

Catalysts	Type of feed	Pressure (psig)	Temperature (^{o}C)	Products observed
HZSM-5	Glycerol	0	400	Carbonyls and Aromatics
		300	300	Carbonyls
		300	400	Carbonyls and Aromatics
	Ethylene glycol	0	300	Cyclic ethers
		0	400	Carbonyls and Aromatics
HZSM-11	Ethylene glycol	0	300	Carbonyls and Aromatics
HY	Glycerol	300	300	Carbonyls
		300	400	Carbonyls and Aromatics
	Ethylene glycol	0	300	Cyclic ethers
		0	400	Carbonyls and Aromatics

Products obtained from one dimensional pore zeolites

Products produced from the conversion of glycerol and ethylene glycol over 20%HMOR, 40%HMOR, and H-OMEGA catalysts under atmospheric pressure are summarized in Table 2. The results show that no aromatics were produced from these three catalysts. Only carbonyls were formed when glycerol was used as feed. However, the 20%MOR catalyst produced cyclic ethers when ethylene glycol was the reactant.

Table 2. Products observed from 20%HMOR, 40%HMOR, and H-OMEGA zeolite catalysts under atmospheric pressure at 300°C and 400 ^{o}C

Catalysts	Type of feed	Temperature (^{o}C)	Products observed
20%HMOR	Glycerol	300	Carbonyls
		400	Carbonyls
	Ethylene glycol	300	Cyclic ethers
		400	Cyclic ethers and Carbonyls
40%HMOR	Glycrol	400	Carbonyls
H-OMEGA	Glycerol	400	Carbonyls

Based on these results, all of the zeolites tested could dehydrate the glycerol. However, to make "gasoline," oligimerization of these products is needed and might be achieved through the structure of the zeolite pores.

HZSM-22 synthesis and characterization

Martens et al. revealed that HZSM-22, a straight channel zeolite with a Si/Al ratio of 53 and a pore diameter of 5.4 °A, was able to oligomerize propylene to trimers with an intermediate degree of branching [Martens et al, 2005]. Therefore, the next step for this study was to synthesize HZSM-22. Figure 1 shows the XRD patterns of the synthesized material compared to the reference XRD patterns of ZSM-22 and ZSM-5 zeolites and reveals that relatively pure ZSM-22 zeolite was successfully synthesized.

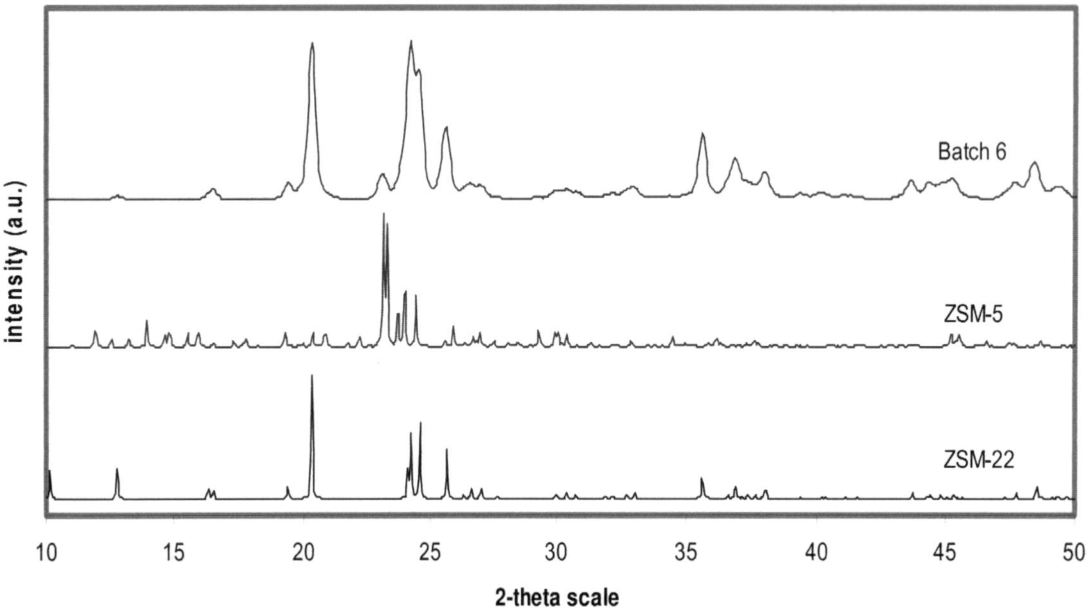

Figure 1 XRD pattern of the synthesized ZSM-22 zeolite compared to the reference XRD patterns of ZSM-22 and ZSM-5.

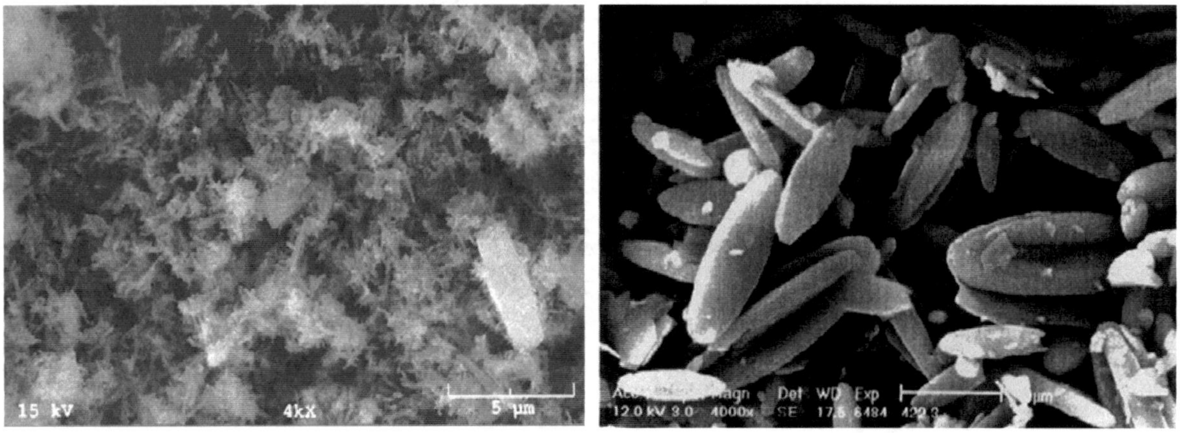

(a) (b)

Figure 2 SEM photographs of (a) the synthesized ZSM-22 zeolites and (b) ZSM-22 zeolites synthesized by Asensi et al.

The SEM photograph of the synthesized ZSM-22 zeolite is shown in Figure 2. It may be observed that the synthesized ZSM-22 crystals are in a needle like shape. Compared to the ZSM-22 synthesized by Asensi et al., it is in the same shape; however, these synthesized ZSM-22 are much smaller. As shown in the Figure 2, the crystal size of the ZSM-22 crystals here is about 1-2 micron. The activity of ZSM-22 zeolite for the conversion of glycerol will be investigated and reported.

CONCLUSIONS

In summary, it can be concluded that the structure of zeolite has the effects on the products as follows; HZSM-5, HZSM-11 and HY produced aromatics, while no aromatics were observed from MOR and OMEGA zeolites. Glycerol may be dehydrated to C_1, C_2, and C_3 carbonyls that might react further to produce fungible gasoline range molecules. However, experiments on the activity of HZSM-22 zeolite and the role of acidity will be carried out to obtain further insight.

ACKNOWLEDGEMENTS

This work has been financially supported by Oklahoma Secretary of Energy and Oklahoma Center for the Advancement of Science and Technology (OCAST).

REFERENCES

Asensi M.A, Corma A., Martinez A., Derewinski M., Krysciak J., Tamhankar S.S, Isomorphous substitution in ZSM-22 zeolite; The role of zeolite acidity and crystal size during the skeletal isomerization of n-butane, App. Catal. A, 174, 163-175, 1998.

Chang C.D., Lang W. H., Smith R.L., The conversion of methanol and other O-compounds to hydrocarbons over zeolite catalysts, J. Catal., 56, 169-173, 1979.

Chu P., Crystalline zeolite ZSM-11, US Pat. 3 709 979, 1973.

Earnst S., Weitkemp J., Martens J.A., Jabobs P., Synthesis and shape selective properties of ZSM-22, App. Catal., 48, 137-148, 1989.

Gayubo A.G., Aguayo A. T., Atutxa A., Aguado R., Bilbao J., Transformation of oxygenate components of biomass pyrolysis oil on a HZSM-5 zeolite. I. Alcohols and Phenols, Ind. Eng. Chem. Rev., 43, 2610-2618, 2004

Gayubo A.G., Aguayo A. T., Atutxa A., Prieto R., Bilbao J., Deactivation of a HZSM-5 zeolite catalyst in the transformation of the aqueous fraction of biomass pyrolysis oil into hydrocarbons, Energy and Fuel, 18, 1640-1647, 2004.

Martens J.A., Verrelst W.H., Mathys G.M., Brown S.H., Jacobs P.A., Tailored catalytic propene trimerization over acidic zeolites with tubular pores, Angew. Chem. Int. Ed., 44, 5687-5690, 2005.

Bio-Ethanol and Btl Diesel Fuel Production from Woody Biomass at Biomass Technology Research Center (Btrc) of Aist, Japan

Kinya Sakanishi, Biomass Technology Research Center (BTRC), National Institute of Advanced Industrial Science and Technology (AIST), Chugoku Center, Kure, Japan

Biomass Technology Research Center (BTRC) of AIST has been conducting the technical developments and organizing the social system for biomass utilization to establish a sustainable society not only in Japan but also in the world in terms of the three important targets, which are described as follows; 1) Ethanol production from woody biomass; 2) Biomass to Liquids(BTL) process, and 3) Study on Biomass Total System. Research activities of BTRC are reviewed in terms of the priority projects on ethanol fermentation from woody biomass through mechanochemical and/or hydrothermal refining pretreatments, BTL(biomass to liquids) by clean gasification, F-T synthesis and catalytic upgrading reactions, and the evaluation of total biomass system by simulation. Especially, feasible utilization technologies of woody biomass resources are most important projects from the viewpoints of CO2 reduction, because their amounts are abundant in Japan and Asian countries, and the woody biomass resources not only contribute to the substitution of fossil fuels such as coal and crude oil, but also considerably reduce the CO2 emission due to their carbon neutrality and deliberate plantation for their sustainable utilization.

HOT SURFACE IGNITION OF ETHANOL, GASOLINE, E-85, DIESEL AND E-DIESEL

Scott Davis and Vijay Somandepalli
Exponent, Inc.
21 Strathmore Road
Natick, MA 01760 U.S.A.
daviss@exponent.com

ABSTRACT

While ignition of flammable and combustible liquids by hot surfaces is a well-known hazard in the chemical processing, automotive and aviation industries, only a limited amount of research has been conducted to understand the ignition mechanism. Recent studies [1-3] have been conducted to better understand this phenomenon for common fluids and fuels in the automotive industry. However, there are very few data on the ignition characteristics of the increasingly popular automotive bio-fuels, such as E-85 and E-diesel. E-85 is composed of 85% ethanol and 15% gasoline, whereas E-diesel is a blend of standard No. 2 diesel fuel containing up to 15% ethanol and a proprietary additive to maintain blend stability and certain fuel properties such as cetane number and lubricity. Currently E-diesel fuels are considered experimental. They can be used legally in off-road applications but for on-road use special permission must be obtained from the U.S. Environmental Protection Agency (EPA). In this paper, we present the results from over 600 ignition tests using a reproducible testing apparatus which includes: a temperature controlled hot plate, a controlled volume of liquid injected onto the hot surface and a quiescent environment. Tests were conducted to study the ignition characteristics of these bio-fuels and compare the results to commonly used gasoline and diesel. Hot surface ignition tests were conducted using 100% ethanol, gasoline (89 Octane), E-85, diesel and E-diesel. Based on the results of the tests, the propensity for the various fuels to ignite on a hot surface was addressed.

INTRODUCTION

While ignition of flammable and combustible liquids by hot surfaces is a well-known hazard in the chemical processing, automotive and aviation industries [1], only a limited amount of research has been conducted to understand the ignition mechanism. Further, there are no known studies documenting this phenomenon in ethanol-blended fuels like E-85, E-diesel and other blends. The ignition of liquids on a hot surface is not trivial and is dependant on many factors, including surface geometry and size, surface material, heat loss, volatility, spray or droplet properties, vapor density, exposure time, fuel structure and vapor or airflow conditions. The general guidance from the NFPA (National Fire Protection Association) on hot surface ignition is that it occurs at temperatures approximately 200°C higher than the auto-ignition temperatures [4].

A few recent studies [1-3] have determined the ignition of gasoline and other hydrocarbons on a hot surface. One study determined the ignition of a variety of automotive and aviation fluids on a horizontal flat plate using a single drop of liquid in a quiescent environment [1]. Their results showed the probabilistic nature of the ignition of fluids on a hot surface and provided a well-defined relationship between surface temperatures and the ignition probability for their apparatus. Another study [2] analyzed the hot surface ignition of 6 different hydrocarbons and showed that the temperature of 50% probability of ignition is strongly correlated with the auto-ignition temperature as has been suggested in the past. They also showed that various blends of gasoline exhibited a weak dependence on their octane ratings and that the volatility of the fuels did not have a noticeable effect on the ignition by hot surfaces. Ref. [3] analyzed the hot surface ignition of gasoline blends on a nodular iron manifold, a stainless steel exhaust manifold and a stainless steel heat shield. This study showed a correlation between the hot surface ignition temperature and the type of metal used and its surface properties. The results for these studies indicate that surface temperature required to ignite gasoline is several hundred degrees higher than the reported auto ignition temperature.

The goal of the present study was to evaluate the propensity of fuels blended with ethanol to ignite on hot surfaces. With the increasing production and use of ethanol blended fuels in the United States, the data presented here will provide valuable guidance towards the development of technologies and processes making use of these fuels in a safe and efficient manner. The data is also hoped to serve as a standalone reference of hot surface ignition data of ethanol blended fuels.

EXPERIMENTAL

Liquid fuels

Hot surface tests were conducted using - 99.5% pure anhydrous ethanol, 89 Octane gasoline and No. 2 diesel – and two ethanol-fuel blends. The ethanol-fuel blends tested were E-85, an 85% ethanol and 15% gasoline blend and E-diesel. The E-diesel was prepared by mixing 7.7% ethanol with 91.38% No. 2 diesel and 0.92% of a proprietary co-solvent from O2-Diesel, Inc., a manufacturer and supplier of E-diesel fuels. The E-diesel mixtures were well mixed prior to each experiment. The gasoline used in these experiments was purchased September 24, 2006 at a Natick, Massachusetts Mobil station and the No 2. Diesel was purchased January 15, 2007 at the same location. The ethanol-blended fuels were mixed and stored in-house at normal room temperature conditions in closed containers. The properties of ethanol, gasoline and diesel are listed in Table 1. The auto-ignition temperature (AIT) for gasoline is sometimes reported as 280 °C as determined by the component with the lowest AIT, however, actual measurements of gasoline blends have yielded AIT of 456 °C. Both values are reported herein. There are no published data available for the ethanol fuel blends and properties of the individual components are commonly used to bound the properties of these mixtures.

Table 1: Properties of Liquids

Fuel	AIT (°C)	BP (°C)	LFL-UFL (% fuel)	Vapor Density (air=1)
ethanol[a]	363	78	3.3-19.0	1.6
gasoline[a]	280,456	38–204	1.4-7.6	3.0-4.0
diesel[b]	257	175-370	0.6-4.7	~5

[a] "NFPA 325: Guide to Fire Hazard Properties of Flammable Liquids, Gases, and Volatile Solids"
[b] Material Safety Data Sheets

Apparatus

The experimental apparatus was designed to accommodate reproducible experimental testing of the ignition of hydrocarbon fuels on a hot surface. The experimental apparatus comprised of two subsystems: A pressurized fluid delivery system and a horizontal, temperature controller hot surface.

The fluid delivery system consists of a nitrogen pressurized fluid reservoir along with a fast acting inline solenoid valve and a spray nozzle, as shown in the cutaway view in Figure 1. The fluid reservoir is mounted to a structure that is insulated to limit the heating of the reservoir and the spray nozzle while conducting experiments. Fluid is sprayed onto the hot surface by pressurizing the reservoir with nitrogen and opening the solenoid valve, thus driving the fluid out an opening on the bottom of the fluid reservoir, past the solenoid valve and through the spray nozzle. The hot surface onto which the fuel liquid is sprayed is an electrically heated 9" x 9" and 0.5" thick square 304-stainless steel plate. The plate is heated via electrical heating elements placed below the plate and controlled by a thermocouple based feedback loop to the heating elements. Complete details of the apparatus and the calibration of the hot surface are given in Ref. [2].

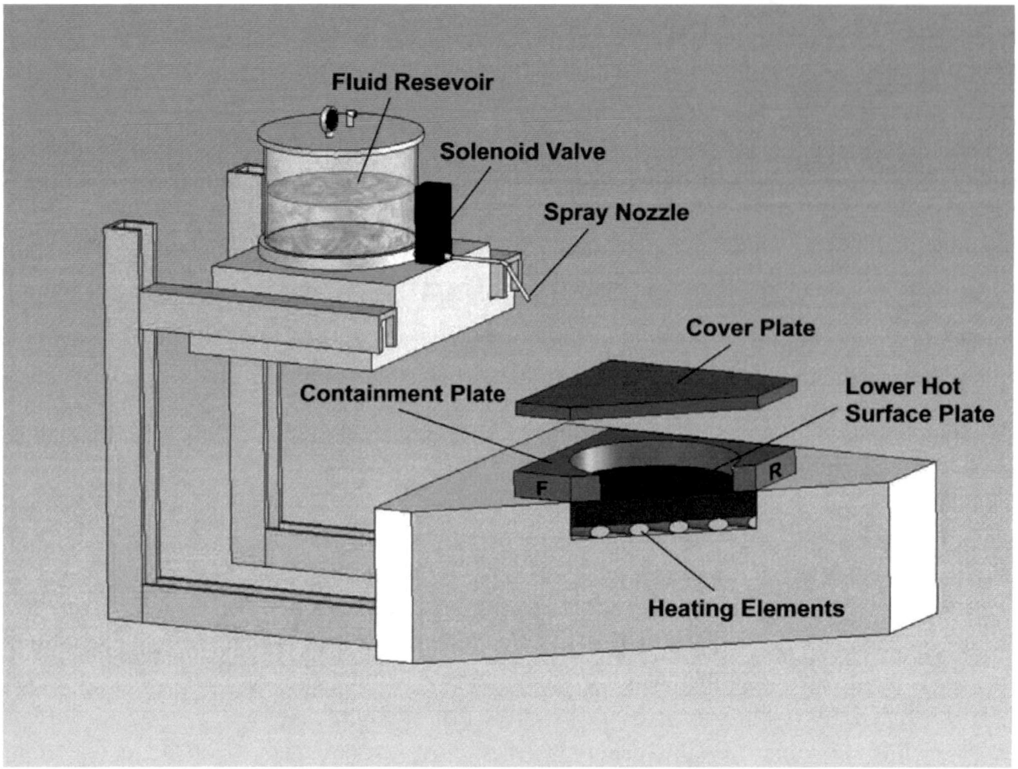

Figure 1: Experimental apparatus

Procedure

The experimental apparatus was used to study the hot surface ignition of the different hydrocarbons identified in Table 1 by spraying a small volume of the hydrocarbon onto the hot surface at various temperatures. Prior to changing fuels after each experiment, the fluid reservoir on the apparatus was emptied and cleaned using ethanol.

For each test event, a constant volume of liquid for each fuel was sprayed onto the hot surface. The amount sprayed per test event as calibrated to be 0.25 ml. This was accomplished by pressurizing the fluid reservoir to 2.5 psig and setting the time duration the solenoid valve was kept open for. Depending on the fuel used, this value ranged from 0.06 – 0.08 seconds. When the spray contacted the hot surface plate, it broke up into multiple droplets on the hot surface.

The experiments were conducted by heating the surface to an initial starting temperature of 250 °C with the cover in place to ensure a uniform surface temperature profile. Once the surface temperature was established, the cover was removed and the liquid sprayed onto the surface. The temperature at the moment of each spray event was recorded along with the resulting ignition/no ignition outcome. The cover was replaced after each spray event to ensure uniform temperatures on the surface. At each set temperature, five separate fuel ejection events were conducted. The temperature of the surface was, then, increased by 10 °C. This procedure was continued until five consecutive ignitions occurred at five consecutive 10 °C increments. Each fuel was tested on at least two separate occasions using the described procedure. This resulted in approximately 10 test points for a given 10 °C temperature range.

RESULTS AND DISCUSSION

For the present study, ignition of the liquids fuels on a hot surface was found to be probabilistic in nature, similar to the findings in Ref. [1]. The experimental ignition probability for ethanol is presented in Figure 2. As observed in Ref. [1], there is a transition regime for which both ignition and non-ignition events can occur. Also shown in Figure 2 is the "S" curve fit of the experimental ignition probability using the equation proposed in Ref. [1]. Fairly good agreement was observed for all fuels between the regression fit and the experimental ignition probability.

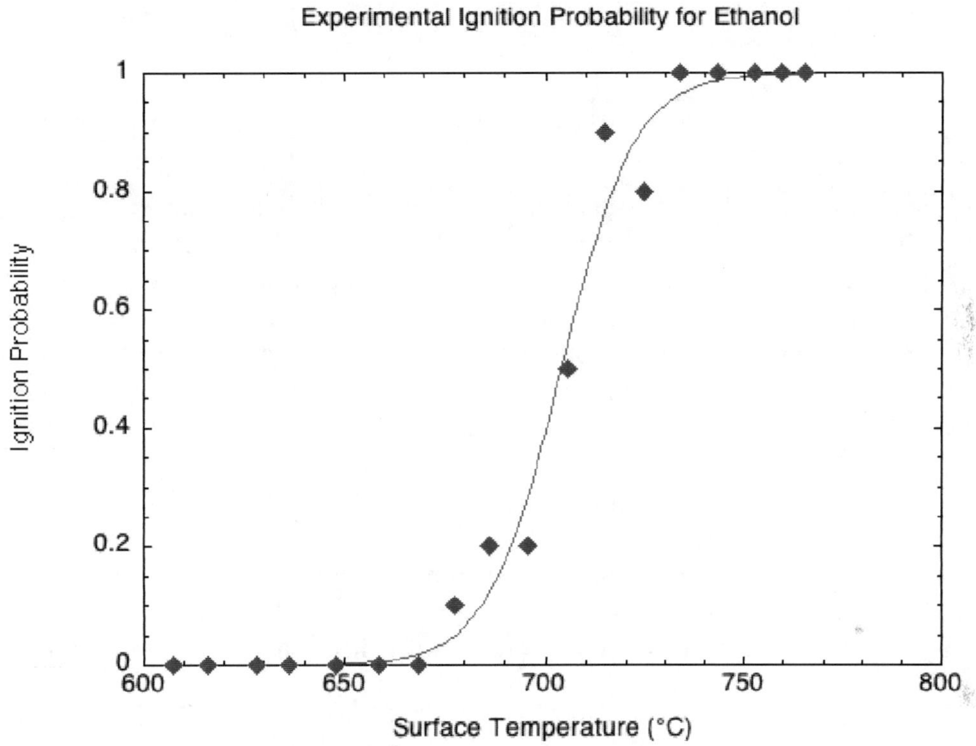

Figure 2: Symbols: Experimental Ignition Probability for Ethanol. Line: Curve Fit of Symbols [1]

Almost all the fuels exhibit a similar behavior of a longer "tail" of low ignition probability and a sharper definition of the temperature for high ignition probability, which may be better described in terms of a skewed "S" curve. Therefore, in the present study we present the temperature limits of this transition regime (i.e. from the lowest temperature that ignition occurs to the highest temperature of no ignition) as well as the 50% ignition probability predicted by the regression curve fit. Results for all the fuels studied can be seen in Table 2.

Table 2: Ignition temperatures

Fuel	Lowest Temp Ignition (°C)	50% Ignition (°C)	Highest Temp No-Ignition (°C)
ethanol	680	704	722
diesel	690	745	755
E-85	705	706	717
E-diesel	685	749	761
89 gasoline	775	802	808
89 gasoline[a]	670	767	785

[a] Results from Ref. [2]

Figure 3 presents our data for 89 octane gasoline, ethanol, diesel, E-85, E-diesel, along with the 89 octane gasoline results from Ref. [2]. For the 89 octane gasoline, the calculated 50% ignition probability and the highest temperature where ignition was not observed agreed well with the results in Ref. [2]. In contrast, the onset of ignition observed in the present experiment, for 89 octane gasoline, was higher by approximately 100 °C than in Ref. [2]. This discrepancy in lower ignition temperature may have been a result of differences in the two gasoline mixtures purchased from different sources and more than a year apart.

The minimum ignition temperatures on a hot surface for ethanol, gasoline and diesel in the present study were all at least 300 °C higher than their reported AIT. E-85 is comprised of 85% ethanol, and as such the ignition properties of E-85 on a hot surface were similar to those of ethanol. Likewise, the E-diesel in the present study is comprised of approximately 91% diesel and had very similar hot surface ignition properties as diesel.

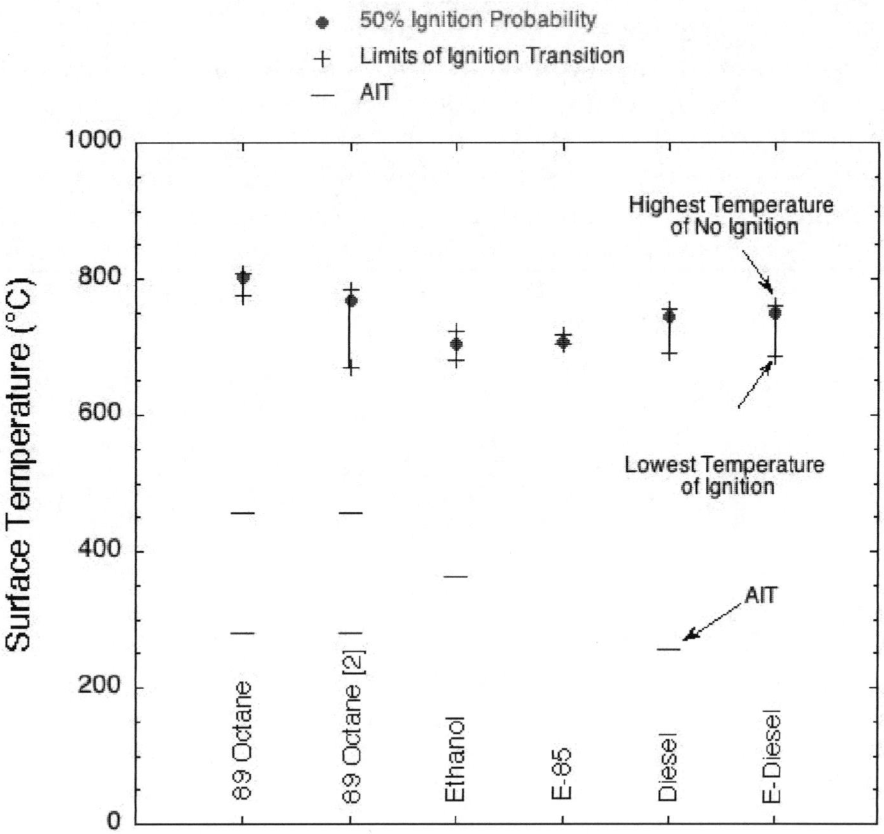

Figure 3: Comparison of Ignition Temperatures on a Hot Surface

CONCLUSION

Hot surface ignition tests were conducted using 100% ethanol, gasoline (89 Octane), E-85, diesel and E-diesel. The minimum ignition temperature measured on a hot surface for ethanol, gasoline and diesel were at least 300 °C higher than their published AIT. Ignition characteristics on a hot surface for E-85 were similar to those of 100% ethanol. Similarly, hot surface ignition temperatures for E-diesel were comparable to those of diesel.

ACKNOWLEDGEMENTS

Exponent would like to thank O2-diesel Inc. for supplying their proprietary additive used to prepare E-diesel. This study would not have been possible without their blend.

REFERENCES:

1. Colwell, J.D., and A. Reza. "Hot Surface Ignition of Automotive and Aviation Fluids," Fire Technology, 41(2):99-117, 2005.
2. Davis, S., Chavez, D. and Kytomaa, H. "Hot Surface Ignition of Flammable and Combustible Liquids," SAE Paper 2006-01-1014, SAE World Congress, Detroit, MI, April 3-6, 2006.
3. LaPointe, N.R., Adams, C.T., and Washington, J, "Autoignition of Gasoline on Hot Surfaces" Fire & Arson Investigator, Oct. 2005: pp. 18 - 21.
4. NFPA 921, Guide to Fire and Explosion Investigations, Quincy, MA 2004 ed.

Ion Exchange Resins for Biodiesel Purification

Daryl J. Gish, Larkin Laboratory, The Dow Chemical Company, Midland, MI 48674

Specialty ion exchange resins can be utilized to bind and remove trace impurities from a biodiesel process stream. After the bulk separation of the glycerin, the biodiesel stream is simply passed though a column of the dried ion exchange resin to bind and remove both ionic salts (such as Na+ and Ca++0, trace catalysts, soaps, and glycerin. The dried resin also acts as a desiccant retaining and removing trace water from the biodiesel feed.

The efficiency and life of a resin to reduce and remove impurities from the biodiesel will depend upon how well bulk separation of the biodiesel and glycerin are made. A resin operation step should be viewed as a polishing process to remove trace materials in the 275-750ppm range, not as a replacement for optimizing the bulk separation.

Generally a lead-lag two column continuous feed system is recommended for most operations. In many cases, a third column should be considered in order to ensure continuous operations and reduce down times. Specialty ion exchange resins, like the DOWEX(TM) DR-G8 resin can be applied to improve and enable a producer to meet ASTM or EN specifications at the point of production to reduce rework and lessen blending concerns.

The Use of a Controlled Cavitation Reactor for Bio-Diesel Production

Douglas G. Mancosky[1], Daniel A. Armstead[1], Tom McGurk[2], Greg Hopkins[2], and Kelly Hudson[1]. (1) Hydro Dynamics, Inc., 8 Redmond Court, Rome, GA 30165, (2) U.S. Boifuels, Inc., 101 Elizabeth Avenue, Rome, GA 30165

Advanced Biofuel Solutions, Inc. of Rome, GA utilizes an innovative next generation reactor technology based upon "controlled cavitation" called the ShockWave Power Reactor (SPR) for process intensification of the transesterification reaction in biodiesel production. The SPR works by taking a feed stock, methanol and catalyst into the machine housing, where it is passed over the generator's spinning cylinder. The specific geometry of cavities in the cylinder and rotational speed create pressure differences within the liquids where tiny bubbles form and collapse. These collapsing bubbles generate shock waves that are used to heat and mix. These bubbles collapse only inside the cavities and away from the metal surfaces and are therefore not damaging to the equipment, the cavitation is controlled. The shock waves increase the surface area of the compounds being mixed so that a higher mass transfer rate occurs. Because of the process intensification, batch processes can be converted to continuous. Cavitation provides a unique mechanism of mixing not found in standard shear devices. The ShockWave Power™ Reactor harnesses cavitation to transesterify vegetable or animal fats in seconds. The process intensity drives the reaction further to completion than with conventional reactors in a fraction of the time. Whether in batch or continuous systems, less reaction time can lead to less saponification or emulsification, allowing you to use a variety of feedstocks with broader range of free fatty acid concentrations than might otherwise be possible.

Commercial Scale Production Data Using the SPR In A Two Stage Reaction System With Poultry Fat

Global View of Technical Aspects of Fuel Ethanol Production

Charles D. Tereck, KATZEN International, Inc., 2300 Wall Street, Suite K, Cincinnati, OH 45212

As the worldwide demand for liquid fuels increases, ethanol production is growing as well and represents the largest bio-based, renewable fuel. The selection of feedstock is discussed as it relates to availability and processing technology employed in different regions around the world. Various technologies and processing schemes are presented that facilitate the efficient conversion of the selected feedstock to ethanol and by-products.

Emerging Technologies for the Production of Biofuels & Biochemicals from Renewables

Ramon Gonzalez, Chem. Engg., RICE, PO Box 1892, Houston, TX 77251

Although biofuels such as biodiesel and bioethanol represent a sustainable, secure, renewable, and environmentally safe alternative to fossil fuels, major scientific and technological breakthroughs are needed for them to become economically viable. Our talk will focus on emerging platforms for the production of biofuels and biochemicals from renewable resources. Special attention will be given to topics relevant to the biological conversion of feedstocks such as cellulosic sugars and crude glycerin along with the integration of fuels and chemicals production (i.e., biorefinery).

Ethanol Production from Mixed Sugars Derived from Lignocellulosic Biomass by the Rite-Bioprocess

Hideaki Yukawa, Microbiology Research Group, Research Institute of Innovative Technology for the Earth (RITE), 9-2 Kizugawadai, Kizu-cho, Soraku-gun, Kyoto, Japan

Worldwide attention is currently focused on bioethanol production from viewpoints of global warming prevention and energy security enforcement. However, feedstock for current bioethanol production processes comprises food crops, which will be in limited supplies in the near future. Therefore, there is a pressing need to use abundant lignocellulosic biomass, some obtained from inedible parts of food crops, is of as demand for bioethanol proliferates.

Corynebacterium glutamicum has widely been used in industrial microbial production of amino acid and nucleic acid. We constructed ethanologenic *C. glutamicum* strains to demonstrate ethanol production under growth-arrested conditions, with cells of the strain packed in a reactor at high density. Growth-arrested conditions were implemented by oxygen deprivation, which enabled the cells to produce ethanol at high volumetric productivity. Now, we are trying to improve the process for ethanol production from mixed sugars containing hexose and pentose sugars derived from lignocellulosic biomass.

This study was partially supported by a grant from New Energy and Industrial Technology Development Organization (NEDO).

Web Page: www.rite.or.jp

Broth Rheology and Ethanol Yield for a *Zymomonas Mobilis* Fermentation with High Substrate Loading

Byung-Hwan Um and Thomas R. Hanley *
Chemical Engineering, Auburn University, 149 Ross Hall, Auburn, AL 36849

Traditionally, as much as 80 percent or more of an ethanol fermentation broth is water that must be removed. This mixture is not only costly to separate, but also produces a large aqueous stream that must then be disposed of or recycled. Integrative approaches to water reduction include increasing the biomass concentration during fermentation.

In this paper experimental results are presented for the rheological behavior of high-solids enzymatic cellulose hydrolysis and ethanol fermentation for biomass conversion using Solka Floc as the model feedstock. The experimental determination of the viscosity, shear stress, and shear rate relationships of the 10 to 20 percent slurry concentrations with constant enzyme concentrations are performed with a variable speed rotational viscometer (2.0 to 200 RPM) at 40°C and combined temperature (50 °C, 30 °C) for the initial 4 hours. The viscosities of enzymatic suspension observed were in range of 0.0418 to 0.0144, 0.233 to 0.0348 and 0.292 to 0.0447 paschal-seconds for shear rates up to $100 s^{-1}$ at 10, 15, and 20 percent initial solids (w/v), respectively. And, the average particle size during the enzymatic treatment and fermentation process of solka floc at 40 and combined temperature (50-30 °C) was approximately 57.8 to70.0 μm, and 44.0 to 57.5 μm for the SSF and SFF process at 10, 15, and 20 percent initial solids (w/v), respectively.

A recombinant strain of *Zymomonas mobilis* (39679:pZB4L) was used in saccharification following the fermentation (SFF) process varying the initial concentration of Solka Floc. The viscosities of fermentation broth observed were in range of 0.024 to 0.028, 0.401 to 0.058, and 0.840 to 0.087 paschal-seconds for shear rates up to $100 s^{-1}$ at 10, 15, and 20 percent initial solids (w/v), respectively. And, the highest ethanol yield, 83.6, 73.4, and 21.8 percent of the theoretical based on the glucose, was obtained at 10, 15, and 20 percent substrate concentration respectively.

The fluid behavior of the suspensions and broth slurries in *Zymomonas mobilis* ethanol fermentation was modeled using the power-law, the Herschel-Bulkley , Cassion, and the Bingham model. The results showed that broth slurries were pseudoplastic with a yield stress. The model slope increased and the model intercept decreased with increasing fermentation time at shear rates normal for the fermentor. The broth slurries exhibited Newtonian behavior at high and low shear rates during initial SFF process.

Ethanol Production from Waste Sweet Potato Using Recombinant Zymomonas Mobilis

M. Nazmul Karim and Binbing Han. Chemical Engineering Department, Texas Tech University, Lubbock, TX 79409-3121

Abstract: This research focuses on converting waste sweet potatoes that are left in the fields and cannot be marketed for various reasons. Sweet potato collected from a field in one of the family-owned farms in Texas is analyzed for its sugar content using HPLC. The sugars in these sweet potato samples include three major components, sucrose, fructose, and glucose. Based on the fresh biomass, the sucrose, glucose and fructose content is 4.35 ± 0.33%, 2.68 ± 0.64% and 2.21 ± 0.43%, respectively. However, since the dry mass content is only 19.4%, the average sugar content based on the dry biomass is close to 50% Prior to the fermentation, the sweet potato was pretreated and the free sugar was extracted using DI water at different temperatures. The results show that the higher the extracting temperature, the higher the extraction efficiency. recombinant Zymomonas mobilis, 8b and AX101 which are obtained from National Renewable Energy Laboratories, Golden, Colorado. Anaerobic fermentation and aerobic incubation have been evaluated and compared in terms of the ethanol production, at various bacteria loads. The effects of the enzyme ratio, the enzymatic reaction time, and the incubation time have been investigated. The results show that recombinant Z mobilis is capable of converting all sugars effectively. Additional studies are designed to optimize the fermentation conditions, and perform a feasibility study for a small industrial scale plant.

Metabolic Engineering of Cyanobacteria for Bioenergy Production

Pengcheng Fu and Jason Dexter. Molecular Biosciences and Bioengineering, University of Hawaii at Manoa, 1955 East-West Road, #218, honolulu, HI 96822

This research addresses one of the pressing national priorities in the United States: seeking bioenergy for the reduction of America's oil addiction. By using fossil fuels as energy sources, we are facing several problems, such as reduced fossil fuel resources; America's dependence on foreign oils; environmental pollutions and the global warming effect, etc. Development of renewable energy is rapidly embraced by our society and industry to meet the energy growth and emission reduction goals. Bioethanol has recently surged to the forefront of renewable fuels technology. It is thus a valuable alternative of fossil fuels. Most of the current ethanol production is from the fermentation of starch crops. This method suffers from some significant obstacles in the establishment of ethanol as a major player in the fuels market. Energy production from agricultural crops has been found to not to be cost-effective and energy-efficient. It also raises environmental concerns due to its emission of hazardous air pollutants and volatile organic compounds, and CO_2 to the environment. The competition of energy vs. food will cause further shortage of the world grain supply and damage to the food security.

We have developed an integrative biological and engineering approach to construct novel cyanobacterium Synechocystis strains for direct conversion of sunlight and CO_2 into ethanol. We have successfully transformed the pyruvate decarboxylase (pdc) and alcohol dehydrogenase II (adh) genes from Zymomonas mobilis into Synechocystis sp. PCC 6803. Our preliminary studies have confirmed that it is feasible to create such ethanol producers from cyanobacterial systems. We are continuing our effort to optimize our ethanol production approach so that it may topple the global dependence on fossil fuels. It is found that the photosynthetic cyanobacteria can be redesigned for highly efficient ethanol production by the combination of gene transformation, strain/process development and metabolic modeling/ profiling analysis. Our goal is to create a sound and sustainable method to reduce the cost of ethanol production to the point that the ethanol production is competitive to utilization of fossil fuels.

Big Energy Savings on a Small Budget: Integrated Use of Process Simulation, Analysis and Synthesis

Rian Reyneke, Advanced Process Design group, AspenTech, 2500 City West Blvd, Houston, TX 77042

High energy prices and tighter environmental regulations are forcing U.S. operators in the petrochemical industry to look at ways to reduce their energy consumption and remain competitive. There are various solutions available in the market to help improve operation of existing facilities such as advanced control and real-time optimization, but these approaches often address only a portion of the overall opportunity.

Fundamental design factors such as plant age, capacity, feedstock and technology employed can play a much more significant role in setting the practical minimum energy target for a unit, so for many the energy gap toward best in class performance can only be bridged through significant design changes. The gap between actual and best in class performance is often very significant, e.g. Solomon data suggest an average EEI for North American Ethylene crackers of 175 (1995 data), compared to plants in Asia at 126 (state-of-the-art = 100).

Despite the size of the opportunity, large-scale process technology upgrades are often not economically feasible due to tight capital payback requirements. The presentation will discuss a more incremental revamp approach where process modeling, analysis and synthesis techniques are applied to identify innovative low capital cost design solutions to help reduce the energy gap toward best in class performance.

Energy Reduction Investment Analysis: Price Volatility Effects

Douglas C. White, Emerson Process Management, 2119 Albans, Houston, TX 77005-1517

Energy prices have experienced significant volatility in recent years. For example, natural gas prices ranged from $4 to $15 per MM BTU's in calendar years 2005 and 2006. Future prices are uncertain but are likely to retain a high level of volatility. This volatility complicates analysis of potential plant capital investments to reduce energy usage, in particular those that involve consideration of alternate energy sources, since traditional financial investment valuation assumes that future cash flows are known exactly. Yet, this is clearly not the case for many energy saving investments. Failure to recognize the effect of volatility on economic analysis can result in incorrectly valuing the potential financial return of the investment. In this paper, appropriate techniques to evaluate such investments are presented along with case studies illustrating the approach.

R&D Opportunities for Energy Efficiency in the Chemical Industry

Dickson E. Ozokwelu, Office of Energy Efficiency & Renewable Energy -- Industrial Technologies Program, US Department of Energy, 1000 Independence Avenue, SW, Washington, DC 20585, Joseph V. Porcelli, JVP International, Inc., 102 Lincoln Street, Staten Island, NY 10314, Peter O. Akinjiola, Psage Research Inc., 234 Village Walk Drive, Macungie, PA 18062, and Joseph Rogers, AIChE, 3 Park Ave., New York, NY 10016-5901.

The U.S. Department of Energy, Office of Energy Efficiency and Renewable Energy's (DOE/EERE), Industrial Technologies Program (ITP) supports research and development (R&D) to improve the energy efficiency and environmental performance of industrial processes. The program's primary role is to invest in high-risk, high-value R&D projects that will reduce industrial energy requirements while stimulating economic productivity, growth, and global economic competitiveness.ITP's Chemicals' subprogram, which supports R&D relevant to the chemical industries, sponsored a study to help guide its research decision-making process and ensure that Federal funds are spent effectively on the kind of projects that support DOE mission and goals. During the study, energy and exergy analysis were applied to 53 highest energy-consuming chemical manufacturing processes to determine the sources of inefficiency and to locate potential process-specific areas for energy recovery. ITP needed to know which manufacturing processes are the most users of high quality energy and therefore the least efficient in consumption of high quality energy. The study provided valuable insights into potential targets for the development and adoption of advanced, energy-efficient technologies in chemicals manufacture, and has become an important tool for assessing future directions in chemicals R&D conducted under the ITP Chemicals subprogram. The inefficiencies of existing technologies and processes in the 53 processes studied were identified and quantified. The locations of energy losses were pinpointed. Recoverable energies for each process were calculated and energy losses in major unit operations that are common across the chemical processes studied were examined. The methodology used for the study as well as the results and trends observed for the 53 process technologies studied will be presented in this paper.

Rotary Kiln Design Based on the Hypothetical Kiln

Louis Herrington, P.E., LEHCO, 112 Wildoak Dr., Daphne, AL 36526

The author previously presented two Papers on the subject of pebble bed regenerative heat transfer. One at the AIChE 2004 Spring Meeting entitled Simplified Regenerative Heat Transfer and one at the 2006 meeting of IT3 entitled Schumann's Curves Replaced by a Simpler Derivation. A third Paper entitled The Hypothetical Kiln was presented at the 2005 Spring Meeting of the AIChE. It was completed late and not published in the Proceedings. It was very early into the discovery of the associated theory and most of the related concepts and equations had not been developed. This Paper shows how the principles of regenerative heat transfer in pebble beds can be applied to Rotary Kilns. The Hypothetical Kiln concept represents a departure from the traditional Rotary Kiln design. For the traditional design method the reader is referred to the Paper "Recover Heat from Hot Solids by Kamal Adham. Dr. Adham described a traditional kiln design that followed from the text by Traybal.

As a brief review of the Author's previous Papers, it was shown that two S-Curves form within a cylinder of pebbles when subjected to a flow of gas at a different temperature. All heat transfer within the cylinder occurs between these S-Curves. In practical terms it was shown that the length of this Transition Region was finite. An equation for the velocity of the Transition Region was derived and showed the velocity was independent of the rate of heat transfer. An equation was also derived for the two S-Curves and stemmed from an analysis of the initial face heating and gas cooling curves. The derived equations were only valid for the case of a uniform temperature pebble bed

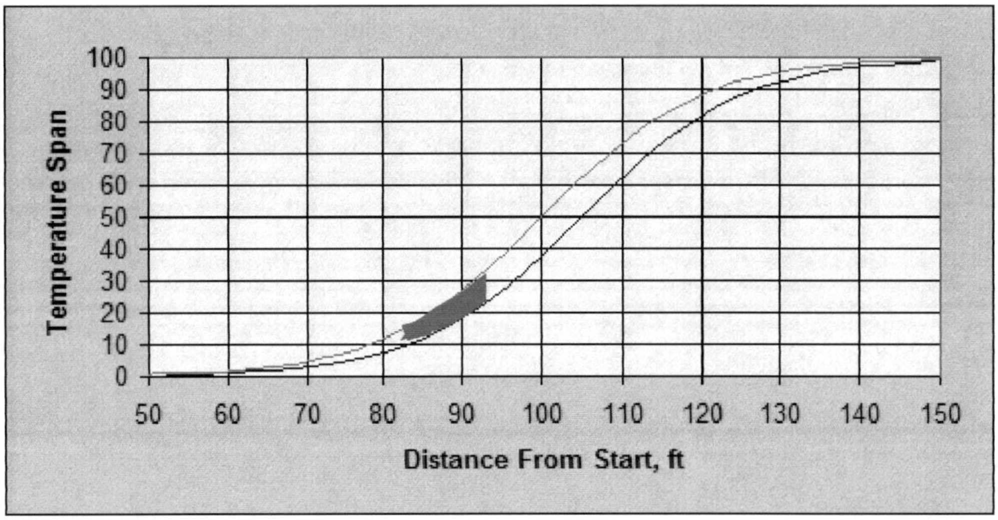

Figure 1. Double S-Curves in a Hypothetical Kiln span a much larger temperature range than the actual Kiln. An actual Kiln is pictured conceptually in red.

The relationship between an actual kiln and the hypothetical kiln is best understood by examining Figure 1. Representing the actual kiln pictorially is the red part of the double s-curves. Consider that the hypothetical temperatures span from 0 to 100%. Actual temperatures of the kiln cover only a portion of the Hypothetical Kiln span. It should be recognized that the double s-curves shown are actual representations of temperatures in an "ideal" pebble bed. In the earlier study of pebble beds by the Author it was stipulated that the s-curves were symmetrical thus adding a new assumption to those in Jacob's Heat Transfer Text. While the assumption of symmetry is not likely to be fully correct it is quite convenient and earlier studies indicated it was true in the ideal case (the heat integral matched the input from the hot gas) and is certain to be a reasonable approximation.

Basis for the design of Rotary Kilns using the concept of a Hypothetical Kiln is that the Hypothetical Kiln can contain a complete S-Curve for the solids and gas. The actual kiln is part of the much longer kiln. As a way to improve understanding it can be said that the Hypothetical Kiln is attached to the double S-Curve moving through an infinite pebble bed. The meaning of a complete S-Curve is that solids are cooled to the hypothetical temperature of the inlet gas and the exit gas is heated to the hypothetical temperature of the inlet solids or visa versa. This is the actual case in a pebble bed. In a Hypothetical Kiln the heat transfer process is completed just as in a pebble bed. The actual kiln occupies only a portion of the two S-Curves. Many properties of the pebble bed apply to the Hypothetical Kiln and become useful in the design, analysis and understanding of the actual kiln.

Understanding Figure 1 will make comprehension of the following much easier. There are numerous equations that describe all aspects of the Rotary Kiln on the basis of the Hypothetical Kiln. However, it is not likely that everything about the relationship has been developed.

There are a number of equations that predict the residence time of pebbles in a kiln. In dryers, residence time equations include the gas velocity. Equations that are generally referred to as "Simple Equations" such as those of Sullivan, et al, Zablotony and the Bureau of Mines, show the residence time as a linear relationship with kiln length. For development of the Hypothetical Kiln concept it is valuable to rearrange these equations to give a characteristic pebble velocity. Then, in a kiln with a fixed rotation and slope only one pebble velocity will apply such as with the Bureau of Mines Equation shown below. Actual burden thickness tends to diminishes through the kiln due to pebble size and velocity segregation unless bridges are placed in the kiln refractory. For derivations and discussions, the burden loading along the kiln will be considered a constant. That will simplify the understanding of principles associated with the Hypothetical Kiln. One last point about Pebble Beds is that the S-Curve for the gas is a perfect reflection of the bed S-Curve. Equation 1 describes the pebble S-Curve and Equation 2 the gas S-Curve for a packed pebble bed. The Rs and Rg symbols are temperature ratios and will be defined later.

- Equation 1 $X = v[\theta-(\sigma C_s/h')Ln(R_s)]$
- Equation 2 $X = v[\theta+\Delta\theta-(\sigma C_s/h')Ln(R_g)$

X	axial distance along pebble bed.
v	velocity of S-Curve through the bed.
θ	total heating time.
$\Delta\theta$	the incremental time that the gas S-Curve precedes the pebbles.
σ	Pebble density.
C_s	specific heat of pebbles.
h'	volumetric heat transfer coefficient. Equal to $h(A_p/V_p)$.

h	heat transfer film coefficient per unit area.
A_p	average surface area of pebbles.
V_p	average volume of pebbles.
R	ratio of temperature differences yet to be defined.
$_s$	subscript for pebbles.
$_g$	subscript for gases.

These same equations when derived for a kiln are altered to reflect the physical relationships within the kiln. The terms $\sigma C_s/h'(1\text{-}e)$ represent the face of the packed pebble bed initially subjected to the cooling gas or in this case air. Air flows directly through the bed but in a kiln the cooling air first encounters the walls and burden and can also encounter pebbles tumbling across the surface of the bed. Since air is a non-absorbing gas for radiation no term for radiation will be needed. New terms must include convection from the wall circumference after correction for being shortened by the burden. If the film coefficient for the burden and the wall are each known, the wall and burden can be handled separately. Otherwise the wall film coefficient will be applied to both. Without showing the geometry within the kiln it can be said that the factor F represents the correction for the length of the wall circumference as shortened by the burden. The factor F is derived in terms of the fraction of radius not occupied by the burden. Then the entering gas will convect heat per foot of wall and burden surface as per $Fh_w piD(Ts\text{-}Tg)$. Also some particles in the kiln can be thrown across the burden surface after being rolled up the side of the kiln. Entering air will contact such particles, if present, in the same way particles are contacted in a pebble bed. Some fraction, f, of the burden can roll and bounce across the surface to the burden's lower end. Candidates for the roll and bounce are the fines that segregate to the floor and some larger pebbles that occupy the surface. Many pebbles tend to roll across others while submerged. Consider, for example, a kiln with a smooth surface filled with roller bearings. The fraction of pebbles rolling and bouncing across the surface could be zero up to the rotation rate where momentum exceeds gravity forces. Conversely, if particle shapes and friction prevent rolling and sliding on the kiln wall and the interior of the burden, the maximum number of pebbles will be rolling and bouncing. Under such extreme conditions the mass rate of pebbles will equal the total burden divided by the time required to move the bottom of the burden to the release position. Release position would represent the point where frictional forces are overcome by gravity and momentum. That position of release can vary and will be a function of the angle of repose, rotation rate and kiln diameter. The second part of that quantity is the travel time across the burden to the kiln bottom. While that quantity might be quantified for the extreme case described above it is somewhat complex and will not be presented. For this treatment the quantity will be represented as a fraction, f, of the burden volume. This treatment will not detract from the validity of the presentation and will simply become a factor to be considered when correlating heat transfer rates for different kilns.

Pebbles in contact with air at the surface transfer extra heat according to $dQ/dX= f\text{Å}(1\text{-}e)h(Ap/Vp)(Ts\text{-}Tg)$. By tradition $h(Ap/Vp)$ is defined as h' the volumetric heat transfer coefficient. When deriving the kiln equation the element that is cooled is the burden and in a kiln is expressed per unit length as $\text{Å}\sigma(1\text{-}e)Cs$. The heat transfer terms are therefore divided into this expression as was shown in Equations 1 & 2. Equation 3 and 4 are the kiln equations equivalent to Equations 1 and 2 for pebble beds. The new terms are defined below.

- Equation 3 $\quad Xs = v[\theta\text{-} (\text{Å}C_s\sigma(1\text{-}e))/((Fh_w piD)+ f\text{Å}h'(1\text{-}e))Ln(R_s)]$

- Equation 4 $Xg = v[\theta+\Delta\theta-(ÅC_s\sigma(1-e)(Fh_w piD)+ fÅh'(1-e))Ln(Rg)]$

e	porosity of bed.
Å	burden cross section.
F	correction for cylinder circumference caused by the burden.
f	fraction of burden in free flight
θ	hypothetical time that kiln has been in passage through the hypothetical pebble bed.
Δθ	lead time for the gas S-Curve.

Equations 3 & 4 are referred to a position at the inlet of a long Hypothetical Pebble Bed through which the Hypothetical Kiln is traveling. The hypothetical time in Equations 3 and 4 refer to the hypothetical beginning of the heating or cooling process. Lead time, Δθ, for the gas is a constant and determinable. The gas S-Curve moves as far ahead of the pebbles as necessary so that heat is transferred. All heat transfer occurs between the s-curves. Under this concept of kiln design the horizontal distance between pebble and gas S-Curves as represented by any particular temperature remains constant. Differences in temperature at any position along the kiln are controlled by heat transfer rate.

Velocity of the S-Curve or Transition Region, (TR), was derived for the pebble bed in the previous Papers. Velocity of the TR is represented by Equation 5. Equation 5 applies when the kiln is in thermal equilibrium and only the pebbles are changing temperature.

- Equation 5 $v = mCg/ÅCs\sigma(1-e)$

m	mass rate of gas through the kiln.

Equation 5 represents the plug flow of heat through the bed and is a very good approximation. The error depends on the point of reference selected. The error is slightly different if the reference is the inlet of the Hypothetical Kiln or if it is the inlet of the actual kiln. For example, if the reference is the actual kiln the denominator in Equation 5 becomes $\rho AgCg + \sigma AsCs(1-e)$. Such an error is not normally significant and for the case of 100 lb/ft3 solids in a 10 foot diameter kiln is less that 0.3% at 1000'C.

The denominator is connected by continuity to the mass flow of solids as shown in Equation 6.

- Equation 6 $Å\sigma(1-e) = w/Vs$

Vs	solids velocity. (Not to be confused with v)
w	mass rate of pebbles.

There are a number of equations that predict the solids velocity in a kiln. Here, the Bureau of Mines Equation as presented in Perry's Handbook and rearranged in Equation 7 is applied.

- Equation 7 $Vs = L/t = kDNS$

L	length of kiln
k	a constant, equal to 315.6 in English units with Vs in ft/hr.

D kiln inside diameter.
N kiln rotations per minute.
S kiln slope in distance of drop per unit distance.

Equation 5 can be combined with Equations 6 & 7 to give velocity, v, in familiar terms as shown in Equation 8.

- Equation 8 $v = (mCgVs/wCs)$

Since the heat transfer in a kiln can be attributed to at least three mechanisms that include radiation it is convenient to express the denominator in Equations 3 & 4 as shown in Equation 9.

- Equation 9 $\hat{H} = FhpiD+f\, h'(w/\sigma Vs)+r$

Now Equation 9 contains a radiation term, r, but it should be recognized that it is difficult to represent radiation as a constant and in reality will increase heat transfer more in the higher temperature zone. In a simplified presentation the radiation "constant" might be expressed by Equation 9A as described in Perry's Handbook.

Equation 9A $r = (\delta\varepsilon 10^8/L)\ [(Ts/100)^4-(Tg/100)^4]/(Ts-Tg)avg$

δ Planks constant
ε` Emissivity

Correlations of kiln heat transfer data should recognize the differences in kiln conditions. Some kiln operations impart less momentum to dry solids and others involve air with no radiant heat transfer except between the walls and pebbles. Each type must be recognized allowing differences to be correlated with momentum or radiation.

Next, define the ratio in Equations 3 & 4 in accordance with Equation 10.

- Equation 10 $\acute{K} = (wCs/Vs\hat{H})$

Now Equations 3 and 4 can be combined with Equation 10 to result in Equations 11 & 12.

- Equation 11 $Xs = v(\theta- \acute{K}Ln(R))$

- Equation 12 $Xg = v(\theta+\Delta\theta- \acute{K}Ln(R))$

The values of Rs and Rg are defined by Equations 13 & 14.

- Equation 13 $Rs = (Tso-Ts)/(Ts-Tgo)$

- Equation 14 $Rg = (Tso-Tg)/(Tg-Tgo)$

Tso hypothetical temperature of the inlet pebbles.

Tgo	hypothetical temperature of the inlet gas.
Ts	actual temperature of the pebbles.
Tg	actual temperature of the gas.

The hypothetical temperatures have to be determined and are found by solving simultaneous equations. The required simultaneous equations will be determined later. Before deriving the new equations a review of the heat and material balance for the Kiln is necessary. Consider the case for cooling hot pebbles. A heat balance allows Equation 15 to be formulated relating the heated gas temperature to the temperature of the exit pebbles. Equation 13 relates any end of kiln pebble temperature to a gas temperature. The designer must determine a suitable value for the end of kiln temperatures.

Equation 15 $$Tg=Tg1 + (WCs/mCg)(Ts1-Ts)$$

Another useful equation represents Ts-Tg as a linear function of Ts as in Equation 14. If mCg/WCs is represented by k then:

Equation 16 $$Ts = [k/(k-1)](Ts-Tg) + (kTg2-Ts2)/(k-1)$$

From a suitable correlation of heat transfer coefficients a value for the heat transfer constant, $\hat{H}$ in btu/hr-°F-ft must be determined. Present correlations of heat transfer are either in terms of volume or square feet. Perry's Handbook list $Us = 23.7(G)^{0.67}$ with Us in units of heat over temperature, time, per square unit of area. From Equation 15 or 16, develop a desirable Log Mean Temperature difference. Then, determine the Kiln length from Equation 15.

Equation 15 $$L = WCs(Ts1-Ts2)/(\hat{H}\Delta Tlm)$$

If the kiln is of known length then simply adjust Ts2 and Log Mean Temperature until the length is satisfied. Gas heat capacity is often set near 1.5 times the mass flow heat capacity. Note that the English units of $\hat{H}$ are btu/hr-°F-ft. In this paper, the coefficient was determined by back calculation.

Derive Equations for Determining Hypothetical Temperature

Step one is to write Equation 11 in two ways. Write it once in terms of the inlet temperature of the pebbles and once in terms of the pebble exit temperature while allowing the hypothetical temperatures to be unknown. If the smaller equation containing Rs1 is subtracted from the equation containing Rs2 the result is Equation 16. Equation 17 is the duplicate of Equation 16 except written in terms of Tg.

Equation 16 $$L = v\acute{K}Ln(Rs2/Rs1)$$

Equation 17 $$L = v\acute{K}Ln(Rg2/Rg1)$$

Note that when the hypothetical temperatures are known, Equations 16 and 17 will predict the temperature of the particles only within the kiln. That is because the distance from the start of the hypothetical bed to the kiln entrance is represented by Equations 11 and 12. Now the point of reference has shifted to the kiln and the median temperature within the kiln represents the plug flow progress of heat flow.

If Rs or Rg is substituted for Rs2 and Rg2 these equations now represent any position along the kiln. Subtracting the revised form of Equation 17 from the same form of Equation 16 at any given temperature will result in Equation 18.

Equation 18 $$Xs-Xg = v\acute{K}Ln(Rg1/Rs1)$$

The final form of Equation 18 results from canceling identical values of Rs2 and Rg2. From Equations 16, 17, and 18 some properties of the Hypothetical Kiln are apparent.

- The ratio R2/R1 can be determined from kiln parameters.
- The ratio Rg/Rs is a constant for any value of Ts.
- Distance between equal values of temperature along the pebble and gas curves is constant.

The ratio Ln(R2/R1) for pebbles or gas is found by rearranging Equations 16 or 17 and combining with Equations 8 and 10 to give the general expression in Equation 19.

Equation 19 $$Ln(R2/R1) = LH/mCg$$

Once the ratio R1/R2 is known, the defining Equations 13 and 14 can be rearranged resulting in two equations with two unknowns. One equation will represent the pebbles and the second to represent the gas. Equations 19 and 20 are the results.

Equation 19 $$R2/R1 = [(Tso-Ts2)(Ts1-Tgo)]/[(Ts2-Tgo)(Tso-Ts1)]$$

Equation 20 $$R2/R1 = [(Tso-Tg2)(Tg1-Tgo)]/[(Tg2-Tgo)(Tso-Ts1)]$$

There are three types of kiln problems to solve.

1. One is for an operating kiln where the objective is to determine the heat transfer constant, $\hat{H}$. Heat Transfer Constant values can be correlated with other similar kilns.
2. A second type is when a Heat Transfer Constant is known or predictable and the capability of an existing kiln for a particular job is to be estimated.
3. A new kiln is to be constructed to meet a particular need.

Some variations exist between the solution of problem 1 and the other two. For example, the first term in Equation 9 is for heat transfer with walls and the bed static surface. For cases with less or insignificant particle momentum and non absorbing gas it is likely that heat transfer results can be predicted from theory and adjusted to reality. Then, if that part of the coefficient is known or predictable, correlations of non-absorbing gas heat transfer constants can be made for the second part of the Heat Transfer Constant represented by the particles in direct contact with gas. Finally, an experimental kiln used with the same conditions of particle momentum and gas flow but using absorbing gases will allow the radiant effect to be determined by difference.

Using the equations derived from the Hypothetical Kiln it is possible to more keenly recognize similarities and differences between operating kilns. Practically all the parameters associated with the kiln appear in the equations. It will no longer be necessary to lump together the results from kilns that have significant differences. For example, particle velocity, kiln rotation, diameter, slope, as well as gas velocity and even the hypothetical temperatures that match up with the particular conditions can be examined for effect on heat transfer. By recognition of the second part of the heat transfer associated with particles in direct contact with gas it should be expected that the influence of particle size as well as particle momentum will be detected. A great deal of experimental work has been done in the past to determine and predict the film coefficients of heat transfer with walls and particles. The correct application of film coefficient predictions might strictly match up with the first part of the heat transfer constant. Then, all observed differences between constants might be related to particles in flight when radiant absorption is not occurring. Differences in heat transfer constants and those predicted using correlated wall film coefficients should be tested against factors related to particle momentum such as rotation, slope, angle of repose, particle size, and any other factors that can influence the release conditions of particles falling from the lifted side of the burden.

There are aspects of the Hypothetical Kiln that have not been sufficiently developed. The relationship of R values and the ratio of gas and pebble heat capacity is not yet fully understood. For example: can the location of the Kiln along the Hypothetical S-Curves be predicted based on the ratio of gas heat capacity to pebble heat capacity per foot. In a separate study it was determined that when the aforementioned ratio was set at one the location became the center of the S-Curve. Applying hypothetical equations to the example by Dr. Adham provides a convenient opportunity to see the new concept. Dr. Adham has kindly agreed to this comparison. Basic data is shown in Table 3.

Table 1. Design parameters for actual case as determined from Dr. Adham's Paper.

		Metric			English	
Pebble flow		10000			22000	
Gas flow		14900	kg/hr		32780.0	lb/hr
Kiln Inside Diameter		2.9	m		9.514	ft
Kiln Length		27	m		88.583	ft
Cross Sectional Area					71.10	ft2
	Hot Pebbles	1000	'C		1832	'F
	Cold Pebbles	150	'C		302	'F
	Cold Gas	29.53	'C		85.15	'F
	Hot Gas	600		1112	1112	'F
Gas flow per unit area		2255.8	kg/m2		461.06	lb/ft2
	Cs	0.25			0.25	btu/lb-'F
	Cg	0.25			0.25	btu/lb-'F
	Qs	2125000			8415000	btu/hr
	Qg	2125001			8415000	btu/hr
Ratio of gas to solids :	k=mCg/wCs				1.490	
A convenient heat balance equation						
Ts=(k/(k-1))(Ts-Tg)+(kTg2-Ts2)/(k-1)						

The case will be treated as an existing kiln. The original Paper was written to shed light on alternative approaches to the cooling of solids. Direct rotary cooling was one type of cooling presented. Only one example of direct Rotary Kiln cooling has been selected for comparison. Basic data is presented below in Table 1. Other tables are presented that cover all the parameters required for design. Velocity of pebbles, velocity of Transition Region, hypothetical hot inlet temperature of pebbles, hypothetical cold inlet temperature of gas, including virtually every parameter associated with a Rotary Kiln is required in the application of the Hypothetical Kiln concept.

Design with a heat transfer coefficient in terms of per foot of length. For this case the value can be determined from a back calculation using Equations 15 & 19.

Table2. Calculation of Ĥ & R ratio by applying Equations 15 & 19

DT1, 'F	217
DT2, 'F	720
DTlm, 'F	419.3
ĤLDTlm=Q, btu/hr	
Ĥ=Q/LDTlm	226.6
Ln(R2/R1)=LĤ/mCg	2.449113
R2/R2=	11.5781

Once the heat transfer constant is known, Equation 19 will determine the R ratio constant. When the value of the R ratio is know it can be used with a rearrangement of Equation 20 to determine the hypothetical inlet and outlet temperatures of the Hypothetical Kiln. This step is shown in Table 3

Table 3. Use kiln parameters to calculate the hypothetical kiln inlet and outlet temperature with a rearranged version of Equation 20.

Ts1		1832
Ts2		302
Tg1		1112
Tg2		85
Ln(R2/R1)=LH/mCg=		2.449113
R2/R1		11.5781
Tgo		Tso
	-71.70	3034
	-71.70	3034

Tso = ((R2/R1)Ts1-Ts2((Ts1-Tgo)/(Ts2-Tgo)))/((R2/R1)-(Ts1-Tgo)/(Ts2-Tgo))
Tso = ((R2/R1)Tg1-Tg2((Tg1-Tgo)/(Tg2-Tgo)))/((R2/R1)-(Tg1-Tgo)/(Tg2-Tgo))

Temperatures of the pebbles along the Rotary Kiln can now be determined from Equations 16 by substituting Rs in place of Rs2. Then L becomes X, the distance along the kiln at the temperature corresponding to the value of Rs. The same is true for Rg in Equation 17. Once a value of Rg at a point in the kiln is known, at one end for example, a review of Equation 18 shows that the ratio Rg/Rs is a constant at any value of X. Taking that advantage can speed the construction of tabulated values for plotting X and kiln temperature. Table 4 shows the constructed table of values for X just in the kiln plus a table of values of x within the Hypothetical Kiln that correspond to values of R and temperature. In order to develop value of x for the Hypothetical Kiln it was necessary to first select an arbitrary value for the length of time the kiln had been operating. When values of Rs are used in the equation for the gas it reveals the time that the gas is leading the solid bed. The gas leads the bed by .5437 hours. It leads a distance of 34.16 ft ahead of the pebbles. With increased familiarity with this method, designs might begin first with the selection of hypothetical parameters.

Data correlations will have to be converted and/or developed based on the new concepts. Better correlations will develop when all kiln parameters are recognized and included. When the back calculated heat transfer constant was compared to the wall film coefficient by dividing by piD, the wall perimeter, it resulted in a value that was quite typical for wall heat transfer of 7.58 btu/hr-ft2-'F. Such a result implies that flying particle influence might not be large. This concept is worth consideration by all those concerned with Rotary Kiln operation and design. The concluding table with graph is shown below.

Table 4. Tabulated values representing the Hypothetical Kiln are in columns 1 & 2. A value of θ = .2647 was selected causing xs to match up with Xs. The column xg shows the corresponding values for the gas.

$mCg/\hat{H} = v\acute{K} =$	36.17	ft		
$Xs1=(mCg/\hat{H})Ln(Rs/Rs1)$				
$Xs2=v[\theta+\acute{K}LnRs]$				
$Xg2=v[\theta+\Delta\theta + \acute{K}LnRs]$			Tso	3034.1
$v=mCgVs/WCs =$	62.83	ft/hr	Tgo	-71.701
$\acute{K}=wCs/(Vs\hat{H}) =$	0.5756	hr		
$Vs =$	42.17	ft/hr	$Tg=(Tgo-RgTgo)/(R+1)$	
$\theta =$	0.2647			
$\Delta\theta =$	0.5437		Rg/Rs =	2.571555

xg	xs	Xs	Ts	Rs	Rs/Rs1	Rg	Tg
122.7	88.58	88.58	302	7.311	11.578	18.800	85.15
120.8	86.64	86.64	320	6.929	10.973	17.818	93.34
117.7	83.57	83.57	350	6.365	10.080	16.368	107.12
113.0	78.84	78.84	400	5.584	8.844	14.360	130.50
74.0	39.81	39.81	1000	1.898	3.006	4.881	456.42
40.5	6.37	6.37	1700	0.753	1.192	1.936	986.00
36.2	2.05	2.05	1790	0.668	1.058	1.718	1070.79
35.7	1.56	1.56	1800	0.659	1.044	1.696	1080.50
34.2	0	0.00	1832	0.631	1.000	1.624	1112.00

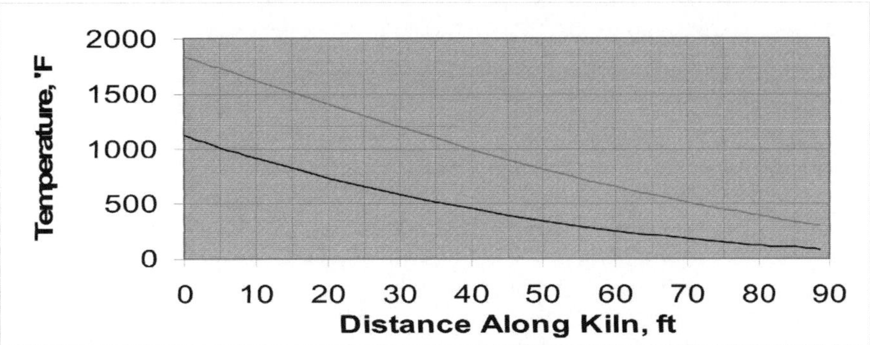

Figure 2. Plot of data in Table 3 showing temperatures along the actual Rotary Kiln.

References

Adham, Kamal, Hatch Associates, "Recover Heat from Hot Solids", Chemical Engineering Progress, August 2000.

Treybal, R.E., Mass Transfer Operations, 3rd Edition, Chapter 12, McGraw-Hill, New York, NY, 1980.

Perry's Chemical Engineers Handbook, Perry, R.H. and Green D.W. 6th Edition, Section 20.

Jakob, Max, "Heat Transfer", Volume II, John Wiley & Sons, New York, NY.

Flow Pattern Simulation in a High Solids Cellulose-to-Ethanol Bioreactor Using Computational Fluid Dynamics

Byung-Hwan Um and Thomas R. Hanley*
Department of Chemical Engineering,
Auburn University, Auburn, Alabama 36849,

Phone: 334-844-7773
Fax: 334-844-2063
Email: hanley@auburn.edu

The agitation system plays an important role in bioethanol production from the Simultaneous Saccharification Fermentation (SSF) and Saccharification Following Fermentation (SFF) processes. To understand and improve mixing and mass transfer in the viscous non-Newtonian systems, flow behavior of fermentation broth was simulated in a bench scale bioreactor (BioFlo 3000). The predictive capabilities of CFD techniques as applied to solid-liquid stirred vessels for a high solids system are investigated.

Based on the angular momentum balance, the required torque in mixing tanks was calculated after the converged solution was obtained as percent of solid suspension. The simulated power number in the turbulent regime was 5.0 for Rushton turbine impeller. And, Reynolds number was 248.6, 64.5, and 50.3 at 10, 15, and 20 percent initial solids (w/v), respectively.

These approaches to simulate mixing tank were comparatively investigated with a baffled Rushton turbine impeller in laminar regime. The standard k-ε turbulent model was employed in the mixing tank simulation with a Rushton turbine impeller. Fluid flow, turbulent kinetic energy and turbulent dissipation rates in three liter have been simulated as solid concentration and RPM. The result visually and analytically showed that there was slow or stagnant flow region which could result in poor nutrient, gas, and heat transfer between top impellers and bottom of the tank.

Heat Exchanger Abstract

Gasketed plate heat exchangers are quickly becoming the most highly used type of heat transfer equipment used in ethanol production today. The ongoing development of plate and gasket material to handle higher temperatures and pressures has allowed gasketed plate heat exchangers to take the place of typical shell and tube applications used in the past. Also, plate heat exchangers can provide average channel spacing up to ¾" to handle high solid fluids. Many plate heat exchangers are being used as condensers, evaporators, and high solids applications in ethanol facilities.

In addition, with many ethanol facilities expanding their production capacity, the flexibility of the plate heat exchanger provides an economic advantage over a typical shell and tube. This flexibility is seen by changing plate configurations to increase velocity or reduce pressure drop or adding plates to expand capacity. Plate heat exchangers provide a smaller footprint than shell and tube heat exchangers and better heat transfer allows for less area needed to perform the same duty.

Fouling and cleaning the plate heat exchangers provide an economic benefit as well. Most plate heat exchangers can be cleaned-in-place or taken apart by 1 or 2 maintenance personnel and reassembled in less than 8 hours. In addition, the small hold up volume, compared to a shell and tube, of the plate heat exchanger provides a savings in the amount of material required to clean the units.

The Production of Hydrogen from Partial Oxidation of Biomass in Subcritical and Supercritical Water - A Thermodynamic Equilibrium Analysis

Shyh-Ming Chern, I-Shou University, Kaohsiung, Taiwan, ROC

Abstract

Hydrogen energy has become increasingly important in recent years because its combustion produces nothing but water vapor and because it is the most desirable feed for fuel cells. It has been demonstrated that high yield of hydrogen gas can be produced from the gasification of biomass materials in supercritical water. However, investigations so far have been mostly experimental, and theoretical studies are scarce. Supercritical water possesses many unique features including high temperature (> 647.4 K), high pressure (> 22 MPa), low hydrogen-bond concentration in addition to other typical characteristics of supercritical fluids, making it a promising, and sometimes ideal, media for the reactions of organics. Despite numerous successful experimental demonstrations, effects of the major operating parameters, e.g. reaction temperature, pressure, water density, feed composition, on the process remain unclear. The present study attempts to investigate the optimum operating conditions of the partial oxidation of biomass materials in supercritical water by means of thermodynamic equilibrium modeling to access the influence of the major parameters. A thermodynamic equilibrium model based on equilibrium equations and equations of state has been proposed, and implemented through an extremely efficient computational scheme that is capable of generating a solution in merely fractions of a millisecond using a notebook computer.

Introduction

Biomass energy is renewable and hence sustainable. The best way of utilizing biomass energy is transforming it into hydrogen energy. The Ideal gasification reaction is:

$$C_6H_{12}O_6 + 6H_2O = 6CO_2 + 12H_2 \quad (1)$$

Previous investigations[1] have shown that gasification of biomass in supercritical water approaches the ideal gasification reaction of biomass, producing dry product gases with higher than 60% hydrogen content. However, most studies to date have focused on gasification of biomass without addition of oxidizing agent, which is quite endothermic and hence is impractical for commercial processes because of the need of massive external energy supply. With proper input of oxidizing agent, the gasification of biomass can proceed with little or no external energy input.

Supercritical water has many unique features including high temperature (> 647.4 K), high pressure (> 22 MPa), low in hydrogen-bond concentration and other typical characteristics of supercritical fluids, making it a promising, and sometimes ideal, media for the reactions of organics. Numerous organic matters have been reacted in supercritical water, e. g. alkanes[2], organic solvents[3], organic wastes[4], glucose, lignin, biomass, etc. However, investigations so far have been mostly experimental, and modeling studies are scarce. In a pioneering work, Tang and Kitagawa[5] modeled the supercritical water gasification of organic materials based on the thermodynamic equilibrium of the effluent and implemented with free energy minimization method. The present study employed a model for the SCWPO of biomass materials, also assuming thermodynamic equilibrium for the effluent. However, the present model is based upon the solution of simultaneous equilibrium equations, rather than the free energy minimization method used by Tang and Kitagawa. For implementation, an equation of state (EoS) is needed to predict the thermodynamic properties of the reaction mixture under various conditions. The Peng-Robinson EOS[6] is employed for this purpose, which was also used in Tang and Kitagawa's work.

Model Formulation

For the reactions of organic materials in supercritical water environment to be kinetically feasible, the reaction temperature has to be higher than 500°C, preferably higher than 600°C. Investigations[6] have shown that thermodynamic equilibrium can be reached in a short period of time as the reaction temperature goes above 700°C. Therefore, thermodynamic equilibrium is assumed in the present model to be attained for the effluents of a SCWPO (or SCW gasification) processes. Organic materials are composed mainly of elements: C, H and O, and thus can be represented by a general formula, CH_mO_n. It has been shown[7] that H_2, CO, CH_4, H_2O and CO_2 are the only chemical species present in significant amount in a C-H-O equilibrium system at relatively high temperature (> 700°C). For thermodynamic analysis, any combinations of independent equations that involves all the above five species will suffice, e.g.,

$$CO_2 + CH_4 = 2H_2 + 2CO \tag{2}$$

$$CO + H_2O = CO_2 + H_2 \tag{3}$$

for which the equilibrium constants, K, can be expressed as functions of activities of species in the mixture, $\hat{a}$, or fugacities of species in the mixture, $\hat{f}$.

$$K_1(T) = \frac{\hat{a}_{H_2}^2 \hat{a}_{CO}^2}{\hat{a}_{CO_2} \hat{a}_{CH_4}} = \frac{\hat{f}_{H_2}^2 \hat{f}_{CO}^2}{\hat{f}_{CO_2} \hat{f}_{CH_4}} \tag{4}$$

$$K_2(T) = \frac{\hat{a}_{CO_2}\hat{a}_{H_2}}{\hat{a}_{CO}\hat{a}_{H_2O}} = \frac{\hat{f}_{CO_2}\hat{f}_{H_2}}{\hat{f}_{CO}\hat{f}_{H_2O}} \quad (5)$$

When ideal solution model (ISM) is adopted, Equations (4) and (5) turn into Equations (6) and (7), respectively, which are written in terms of pure component fugacities and molar fractions.

$$K_1(T) = \frac{f_{H_2}^2(T,P)y_{H_2}^2 f_{CO}^2(T,P)y_{CO}^2}{f_{CO_2}(T,P)y_{CO_2} f_{CH_4}(T,P)y_{CH_4}} \quad (6)$$

$$K_2(T) = \frac{f_{CO_2}(T,P)y_{CO_2} f_{H_2}(T,P)y_{H_2}}{f_{CO}(T,P)y_{CO} f_{H_2O}(T,P)y_{H_2O}} \quad (7)$$

Equations (6) and (7) can be further managed for easier implementation into Equations (8) and (9), respectively.

$$K_1^{'}(T,P) = \frac{K_1 f_{CO_2}(T,P) f_{CH_4}(T,P)}{f_{H_2}^2(T,P) f_{CO}^2(T,P)} = \frac{K_1 \phi_{CO_2}(T,P)\phi_{CH_4}(T,P)}{P^2\phi_{H_2}^2(T,P)\phi_{CO}^2(T,P)} = \frac{y_{H_2}^2 y_{CO}^2}{y_{CO_2} y_{CH_4}} \quad (8)$$

$$K_2^{'}(T,P) = \frac{K_2 f_{CO}(T,P) f_{H_2O}^2(T,P)}{f_{CO_2}(T,P) f_{H_2}^2(T,P)} = \frac{K_2 \phi_{CO}(T,P)\phi_{H_2O}^2(T,P)}{P\phi_{CO_2}(T,P)\phi_{H_2}^2(T,P)} = \frac{y_{H_2}^2 y_{CO}^2}{y_{CO_2} y_{CH_4}} \quad (9)$$

For the abovementioned single-phase system at equilibrium at a specified T and P, there are five independent variables, y_{H_2}, y_{CO}, y_{CH_4}, y_{H_2O} and y_{CO_2}, and therefore, in addition to

$$\sum y_i = 1 \quad (10)$$

two more constraints are needed to uniquely define the state of the system. Any two ratios of the following three elemental balances will suffice.

$$n_{C,feed} = n_{CH_4} + n_{CO} + n_{CO_2} \quad (11)$$

$$n_{H,feed} = 4(n_{CH_4}) + 2(n_{H_2}) + 2(n_{H_2O}) \quad (12)$$

$$n_{O,feed} = n_{H_2O} + n_{CO} + 2(n_{CO_2}) \quad (13)$$

The implementation of the model needs an equation of state (EOS) capable of estimating the fugacities with acceptable accuracy for the five gaseous species in the supercritical water mixture. The Peng-Robinson EoS has recently been evaluated and fine tuned[8], and shown to predict thermodynamic properties of water in the supercritical region with reasonable accuracy, except for water in the immediate vicinity of critical point. The Peng-Robinson EoS[6] takes the form

$$P = \frac{RT}{v-b} - \frac{a(T)}{v(v+b)+b(v-b)} \quad (14)$$

where

$$a(T) = 0.45724 \frac{R^2 T_C^2}{P_C} \alpha(T, \omega), \qquad \sqrt{\alpha} = 1 + \kappa\left(1 - \sqrt{T/T_C}\right),$$

$$\kappa = 0.37464 + 1.54226\omega - 0.26992\omega^2$$

which was originally proposed for nonpolar species and has been well developed for a wide variety of pure components and mixtures. Since PR-EoS was not originally proposed for polar species, significant prediction errors are expected for water. It has been suggested[8] that using $\omega = 0.5491$, instead of the experimental value $\omega = 0.344$, improves the performance of PR-EoS with subcritical and supercritical water dramatically. The calculation of thermodynamic properties of pure components and mixtures with PR-EoS are well documented[6].

Model Implementation and Comparisons

With a known feed composition, Equations (8) through (13) are solved simultaneously for the equilibrium composition at a specified temperature and pressure if ISM for all species is assumed. Pure component fugacities in Equations (8) and (9) are estimated using PR-EoS. Equilibrium constants, K_1 and K_2, as functions of temperature are calculated using standard free energies of formation as well as ideal-gas heat capacity functions of temperature. A special semi-analytical solution procedure, well documented elsewhere[9], has been adopted to solve simultaneous Equations (8) through (13). The ISM solution is then serves as the first guess for the real solution without ISM, and partial fugacity coefficients, $\hat{\phi}_i$, are then calculated from the first guess and replace the pure-component fugacity coefficients, ϕ_i, in Equations (8) and (9). Then a new solution is obtained based on the newly calculated $\hat{\phi}_i$ through the same procedure. The entire procedure is repeated until the improvement of the newest solution from the last one diminishes to an acceptable margin. An average of 3 to 4 iterations is needed for a single set of solution. The solution procedure has been thoroughly tested and proven to be robust as well. In all cases, there exists only one set of solution satisfying all the equations and natural constraints, $0 \le y_i \le 1$.

Reliable experimental data for the binary interaction parameters, k_{ij}, in PR-EoS[6] for mixtures are difficult to obtain. Hence, a sensitivity test has been performed that shows a change of all k_{ij}'s from 0 to 0.1 alters the final solution of gas molar fractions by only about 1%. Therefore all k_{ij}'s take zero value for all computations.

In contrast to Tang and kitagawa's[5] free energy minimization method, which often requires more expertise to apply and is error-prone, and needs more computing resources, the present model combined with the dedicated solution procedure is highly efficient and easy to implement. A computer program coded in Fortran was used to implement the model and facilitate the computations. A single set of solution is obtained within fractions microseconds on a notebook computer with a clock of 1 GHz or about 2000 sets of solution can be obtained in a single second.

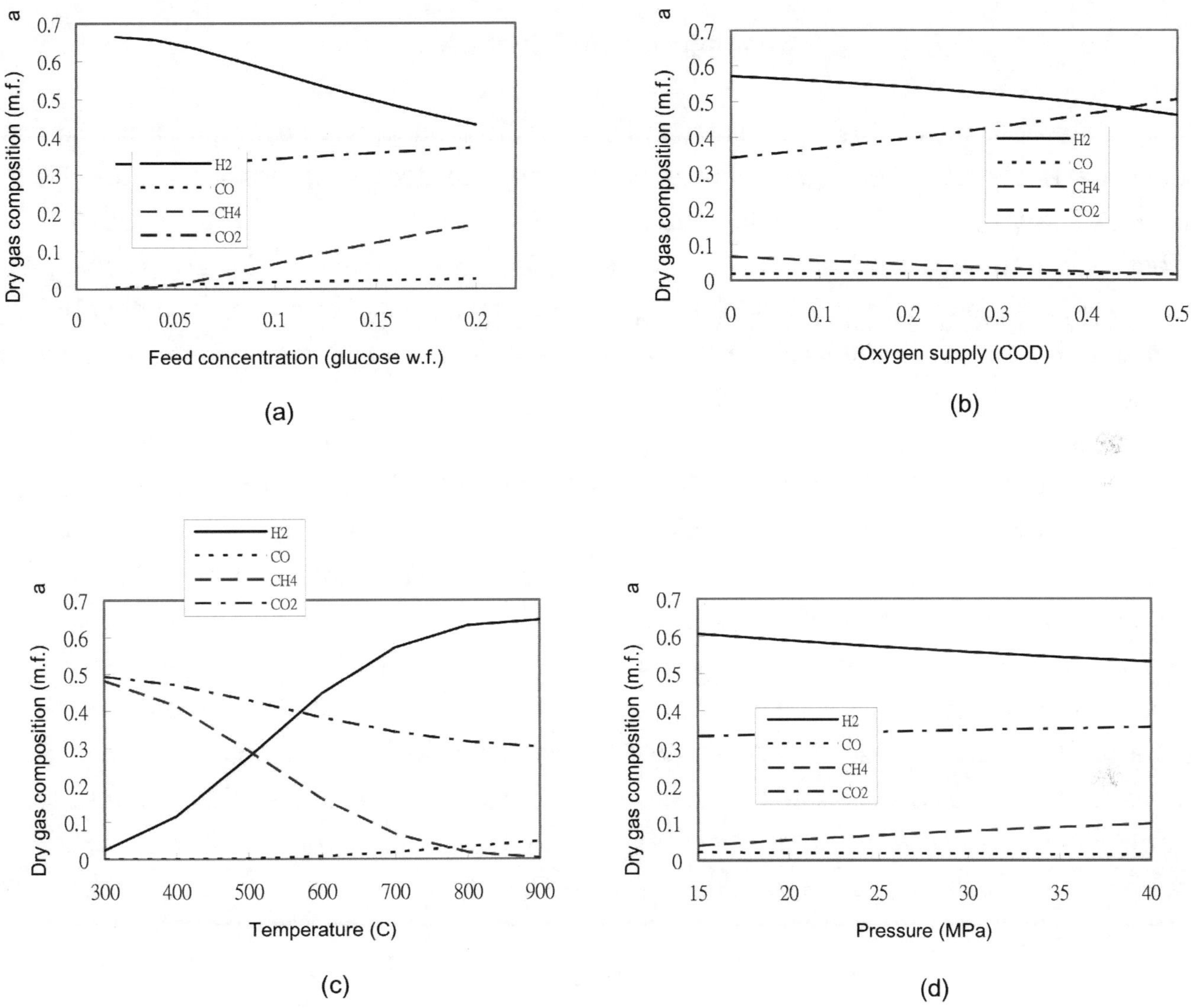

Figure 1. Simulation of SCWPO of glucose, (a) Parameter: feed concentration (700°C, 25 MPa, no oxygen supply), (b) Parameter: oxygen supply (0.1 glucose w.f., 700°C, 25 MPa), (c) Parameter: temperature (0.1 glucose w.f., 25 MPa, no oxygen supply), (d) Parameter: pressure (0.1 glucose w.f., 700°C, no oxygen supply).

In a pioneering work, Tang and Kitagawa[5] simulated the gasification of organic materials by assuming thermodynamic equilibrium with free energy minimization method. They simulation results were compared to the experimental works from literature, including the supercritical water reforming of methanol, supercritical water gasification of glucose, cellulose and real biomass. In general, moderate success was achieved. The key difference between the present model and Tang and Kitagawa's[5] lies in the model formulation and solution method, but the results are essentially identical.

Simulation SCWPO of Glucose

To assess the optimal operating condition for the production of hydrogen from biomass in subcritical and supercritical water, simulations of partial oxidation of glucose in supercritical water are performed and the results presented in Figure 1. Typical values employed for operating parameters are T = 700°C, P = 25 MPa, feed concentration = 0.1 w.f. glucose, and zero oxygen supply, for the simulations. Oxygen supply are measured as the number of chemical oxygen demand (COD), which is equivalent to the stoichiometric amount of oxygen needed to completely oxidize the feed.

The noticeable trends seen from Figure 1 (a) are the decrease in H_2 concentration and the increase of CH_4 concentration in the effluent as the weight fraction of glucose increases from 0.02 to 0.2, while CO concentration remains low. Figure 1 (b) shows that the increase of oxygen supply has only a moderate effect on the composition of the effluent with a slow decreasing trend of H_2 and CH_4. It is significant since the incorporation of oxygen can reduce or even eliminate the energy demand needed to run a SCWPO system, while keeping the adverse effect to a minimum. Figure 1 (c) once again confirms the general observation that temperature is the most influential parameter for a SCWPO system. From 300 to 900°C, the H_2 content increases dramatically from a few percents to about 65% and the CH_4 content reduces to insignificant level, while CO content increases slowly. On the other hand, Figure 1 (b) confirms that pressure effect plays a minor role in a SCWPO system

Conclusions

A model for the SCWPO of organic materials on the basis of thermodynamic equilibrium equations and equations of state has been developed and applied to the production of hydrogen from the gasification of biomass in SCW. The model is implemented through an extremely efficient computational scheme that is capable of generating a solution in merely fractions of a millisecond using a notebook computer. The simulation results indicate that the hydrogen content in the dry product gas increases significantly with decreasing concentration of glucose in feed and with increasing reaction temperature, and is only mildly reduced by the increase of oxygen supply and the increase of pressure. The results also confirms that the hydrogen content in the dry product gas from the SCWPO of biomass could be as high as 70% and the most influential parameter is the reaction temperature.

Acknowledgements

It is grateful for the financial support from I-Shou University (ISU95- 02-11).

References

1. Schmieder, H., Abeln, J., Boukis, N., Dinjus, E., Kruse, A., Kluth, M., Petrich, G., Sadri, E., and Schacht, M., *J. of Supercritical Fluids*, **17**, 2000, 145-15
2. Watanabe, M., Inomata, H., Osada, M., Sato, T., Adschiri, T., and Arai, K., *Fuel*, **82**, 5, 2003, 545-552
3. Kim, Y. L., Chung, S. J., Kim, J. D., Lim, J. S., Lee, Y. W. and Yi, S. C., *Reaction Kinetics and Catalysis Letters*, **77**, 1, 2002, 35-42
4. Park, Y. K., Reaves, J. T., Curtis, C. W. and Roberts, C. B., *Journal of Elastomers and Plastics*, **31**, 2, 1999, 162-179
5. Tang, H. and Kitagawa, K., *Chemical Engineering Journal*, **106**, 2005, 261-267
6. Peng, D. Y. and Robinson, D. B., *Ind. & Eng. Chem. Fund.*, **15**, 1, 1976, 59-64
7. Lee, I. and Kim, M., "Gasification of glucose in supercritical water," *Ind. Eng. Chem. Res.*, **41**, 2002, 1182-1188.
8. Chern, S. M., "Evaluating and Fine Tuning Cubic Equations of State For Supercritical Water Processes," 2007 Spring National Meeting CD-ROM proceedings, New York, NY: AIChE, 2007
9. Chern, S. M., Equilibrium and kinetic modeling of Co-Current (Downdraft) Moving-Bed Biomass Gasifiers, A Ph.D. dissertation, 1989, Chap. 4.

ENERGY AND COST REQUIREMENT TO PRODUCE RICE HUSK BRIQUETTE USING DIFFERENT SCREWS AND HEATING SYSTEMS

M.A.Momin and M.Alam

ABSTRACT

A research was conducted to determine the energy and cost requirement to produce rice husk briquettes (RHB) using different screws (new screw, locally made screw and screw repaired by welding) and die heating systems (electric heating coil made by pipe, kerosene burner, and electric heating coil made by puti). The production rates of RHB were determined at every hour by operating different combination of screw and heating system. The energy consumption and cost were also calculated for producing briquette. The production rate was the highest (110 kg/hr) and the cost was the lowest (1.69 Tk./kg) for repaired screw and pipe coil heating system. It was also observed that the life of screw was in the highest (8-10 hours) position in case of repaired screw. It was also observed that the energy consumption was the lowest (13.55 kW-hr) position in case of repaired screw and pipe coil die heating system. It was in the highest position (17.42 kW-hr) when the heating system was kerosene burner. Therefore, based on the lowest energy consumption and cost to produce rice husk briquette, the best suitable combination to produce RHB was repaired screw and electric pipe coil heating system.

Keywords: rice husk briquette, energy and cost

INTRODUCTION

Rice husk briquetting (RHB) technology is the technique to produce briquette using rice husk as raw materials. This technology was started first at U.S.A in 1865 (Bhattacharya & Shrestha, 1990). Hazi Hayat Ullah proprietor of Agroni Auto Rice Mill, Sylhet, first brought a RHB machine in Bangladesh from Taiwan in 1987. Then, the machines are gaining popularity and showing a high potential use of abundant rice husk to produce an alternative fuel (Briquettee). This briquette can be used in households, hotels, industries, foundry shops, brick field etc. In this process, screw forces rice husk through a heating die. Usually die heating is done by electrical energy. The concept of heating to the biomass for softens the lignin of biomass. This is one of the major components to make bond so there is no need to add any additional binding material to produce briquette. Moral (2001) conducted a survey on present status of briquette machine in Bangladesh and they found that Bangladesh have only 906 machines, in which Dhaka division has only 47 machines. He also estimated that 18000 briquette machines could be run by using only 10-20% of existing available raw materials. That means, a large number of rice-mill owners do not know the technology. Miah et. al. (1999) conducted a survey in Bangladesh and reported that briquette machine was promising and profitable. Alam (2002) reported that number of mill owners were not interested to install a RHB machine due to high electric energy consumption in die heating. The rice mill owners are interested to have a briquette machine and would like to use briquette in the mill, if they get technology with low energy consumption, and training facilities from any source. Therefore, a research was undertaken to find a system that would be required less energy and cost to produce rice husk briquette.

MATERIALS AND METHODS

A traditional RHB machine was operated by an electric motor. The photographic view and cross sectional view of the machine is shown in Fig.1 and Fig.2 respectively. The die was heated by an electric coil (Fig.3). In this study, alternate heating coil (Fig.4) and kerosene burner (Fig.5) were used to produce the RHB. The energy requirement using heating coil and kerosene burners were compared with that for using electric heating coil.

A screw made of mild steel (Fig.6) welded by electrode after 6-8 hours of operation to produce RHB. Initially the screw was manufactured in a lathe machine and the cost of the screw was Tk.600. Another screw (Fig.7) made in local workshop, was also used to produce RHB. For producing the screw in local workshop, mechanics made a taper in a shaft and marked a line on it as required pitch of normal screw. After that, they filled up on the line with a electrode of ferro speed or hard craft. Finally, they finished the screw with the help of a grinder. The cost of locally made screw varied between Tk.200-250 only.

The combination of different screws (new, repaired and locally made) and different die heating systems (puti coil, pipe coil and kerosene burner) were used in this experiment to find the energy consumption, life of screw, production rate, total production and cost per kilogram of RHB.

The hourly productions of briquette using different techniques were measured over the entire working life the screw. The results of production rate were calculated from the average of three trials for each method. A comparative bar graph was then prepared showing the difference of the hourly production rate for using different system.

The energy consumption during the production of RHB was measured by a wattmeter. A stopwatch was used to record the operating time. Then, the energy consumption was determined by the following formula.

Energy consumption = W.t (kW-hr)

Where,

W = Watt meter reading (kW)

T = Operation time (hour)

Physical properties (density and crushing strength) of the briquettes produced by different methods were determined in the laboratory of Farm Structure. The samples of RHB of 100 mm length were collected at the 3rd operating hour of the screw life. A scale was used for measuring the length of RHB and the samples were cut by a power saw.The inside and outside diameter of the RHB samples were measured by an inside calipers, outside calipers and a scale. The cross sectional area of the samples was calculated using the above measurements by the following equation.

$$\text{Cross sectional area of briquette (A)} = \frac{3.14 \text{ X } (D^2 - d^2)}{4}$$

Where,

D = Outside diameter (mm)

d = Inside diameter (mm)

The weights of briquettes were measured by a balance and the densities of the samples were determined by using the following equation. The average densities of RHB were plotted against the time of production of operation to compare the density and the results were converted into gm/cm^3.

$$\text{Density,} \quad \rho = \frac{W}{A \text{ X } L}$$

Where,

W = Weight (gm)

A = Cross sectional area (mm^2) L = Length (mm)

The RHB sample of 100mm size from different methods were crushed by the Universal Testing Machine and readings of load were recorded. The crushing strengths of the samples were determined by using the following equation. The average crashing strength was calculated for each group.

$$\text{Crushing strength } (kg/mm^2) = \frac{P}{A}$$

Where,

P = Load (kg)

A = Cross sectional area (mm^2)

RESULT AND DISCUSSIONS

Production rate of briquette for different screws and heating systems:

The production rate of briquette for a new screw at different hours for different heating systems is presented in Fig.8. It was shown that the production rate of pipe coil was in the highest position (85.33 kg/hr) after the first hour of operation and the lowest production rate was 74.33 kg/hr for kerosene burner. The production rates were gradually decreased with the increase of operating time. At the end of 3 hours of operation, the production of RHB were 62.67 kg/hr and 63.33 kg/hr using puti and pipe coils respectively. That means, the productions were not varied remarkably for pipe and puti coil heating system. In case of kerosene burner, it was 61 kg/hr which was 3.82% less than the production rate from a pipe coil. Therefore, the combination of a new screw and pipe coil die heating system would give the highest production.

The production rates of briquette for a repaired screw at every hour were varied for different heating systems and it is shown in Fig.9. The production rates of RHB using puti coil and pipe coil were very similar except production after 1 hour of operation. It was also observed that the production rate of briquette using kerosene burner was in the lowest position (49 kg/hr). The production rates of RHB using kerosene burner were 31.6% and 28.2% lower than the production from pipe coil and puti coil respectively. At the end of 10 hours of operation, the production rate of RHB using kerosene burner was 19.8% lower than the production rate of briquette using puti and pipe coil heating systems.

The hourly variation of production rate for locally made screw for various heating systems is shown in Fig.10. The production rates were decreased slowly with the time of operations. The production rate was very similar for puti and pipe coil heating systems. However, there was a considerable difference of production rates between electric die heating (made by puti & pipe coil) and kerosene burner heating system. The production rates in all type of heating systems were decreased with time of operation because of continuous wear of screw. It was also observed that the production rate of RHB using a new screw for kerosene burner heating system was less than the production rate of briquette using puti and pipe coil die heating systems.

The total production of briquette were varied for new, repair and locally made screws using different heating systems and shown in Table1. In case of pipe coil heating system and repaired screw, the total production of briquette was in the highest position 838 kg. It was higher than the production from any other combinations of screws and heating systems.

Table 1: Variation in total production of RHB for different combination of screws and heating systems

Type of screw	Heating system	Total production (kg)
New screw	Puti coil	223.67
	Pipe coil	225.33
	Kerosene burner	202.33
Repaired screw	Puti coil	834.33
	Pipe coil	838.00
	Kerosene burner	637.0
Locally made screw	Puti coil	499.67
	Pipe coil	509.67
	Kerosene burner	393.67

Energy consumption to produce briquette for different screws and heating systems:

The average energy consumption producing RHB using kerosene burner was 17% higher than that for using puti and pipe coil (Fig.11). It was also observed that the energy consumptions for producing RHB were very similar using puti and pipe coils and these were 14.48 kW-hr and 14.52 kW-hr respectively. The variation of energy consumption producing briquette using repaired screw for different heating systems is presented in Fig.12. There was nominal effect on energy consumption to produce RHB using repaired screw with puti and pipe coil heating systems. However, the energy consumption to produce briquette using kerosene burner die heating system was higher than those from others die heating systems. The energy requirement for producing briquette using locally made screw and different die heating system is presented in Fig. 13. From this figure, it was found that the energy consumption to produce RHB for kerosene burner was higher than that for puti and pipe coil die heating systems. The energy consumption of different combination of screws and heating systems is shown in Table2.

Table 2: Variation in energy consumption of different combination of screws and heating systems

Type of screw	Heating system	Energy consumption (kW-hr)
New screw	Puti coil	14.48
	Pipe coil	14.52
	Kerosene burner	17.42
Repaired screw	Puti coil	13.82
	Pipe coil	13.55
	Kerosene burner	16.72
Locally made screw	Puti coil	14.67
	Pipe coil	14.79
	Kerosene burner	17.41

Physical parameters of briquette:

The density of briquette producing by different screws and die heating systems is shown in Fig.14. The maximum density of briquette was 1.15 (gm/cc) producing from the repaired screw with pipe coil die heating system. The minimum density of briquette was 1.1 (gm/cc) getting from locally made screw with puti coil and kerosene burner die heating system. The crushing strength of 100mm long briquette producing by different screws with different die heating systems is presented in Fig.15. The maximum crushing strength of briquette was 10.62 Pascal when produced from a repaired screw and pipe coil heating system. The minimum crushing strength of RHB was 10.41 Pascal for producing by using locally made screw with the puti coil heating system.

Life of screw:

The variation of working life of different screws is shown in Fig.16. The life of the new screw, repaired screw and locally made screw were 3, 10 and 6 hours respectively. The working life of locally made screw was 40% less than the life of the repaired screw. The life of repaired screw was higher because of hard craft welding and less erosive.

Production cost to produce briquette for different screws and heating systems:

The production cost of RHB using a new screw for different heating systems is shown in Fig.17. It was the lowest (4.48 Tk./kg) for using pipe coil. The production cost of RHB using kerosene burner was 8.5% and 2.6% higher than the cost of RHB produced by puti and pipe coil die heating respectively. In case of new screw, the production cost was very high because of short life of screw. The variation of production cost with a repaired screw for different heating systems is shown in Fig.18. It was in the lowest position (1.83 Tk./kg) when the heating system is pipe coil. The production cost of RBH using kerosene burner was 15% and 16% higher than the puti and pipe coil respectively. The production cost of different heating systems when operating with a locally made screw is shown in Fig.19. It was found that the production cost was the lowest (2.49 Tk/kg) in case of pipe coil heating system. The production cost of RHB using kerosene burner were 16% and 17% higher than that from the puti and pipe coil respectively. The profit of RHB producing from different combination of screws and heating systems is shown in Table 3. The average selling price of RHB was Tk.3/kg. It was found that RHB produced from repaired screw and pipe coil heating system gave the highest profit (Tk1.17/kg).

Table 3: Variation of profit for selling RHB in different rates producing from different combination of screws and heating systems

Type of screw	Heating system	Production cost (Tk/kg)	Profit on average selling price (3 Tk/kg)
New screw	Puti coil	4.51	-1.51
	Pipe coil	4.48	-1.48
	Kerosene burner	4.93	-1.93
Repaired screw	Puti coil	1.85	1.15
	Pipe coil	1.83	1.17
	Kerosene burner	2.18	0.82
Locally made screw	Puti coil	2.52	0.48
	Pipe coil	2.49	0.51
	Kerosene burner	3	0

CONCLUTIONS

The results revealed that the highest total production was 838 kg for the combination of repaired screw and pipe coil and the lowest production was 202.33kg for the combination of new screw and puti coil. It was found that the lowest energy consumption was 13.55 kW-hr for the combination of repaired screw and pipe coil. It also found that the maximum profit obtained in case of repaired screw and pipe coil heating system. Therefore, it can be concluded that the best suitable combination was repaired screw and electric pipe coil heating system based on the lowest energy consumption and cost to produce rice husk briquette.

REFERENCES

Battacharya, S.C. and Sherestha, R.M. and Sivasakthy, S.1988. State of the Art of Biocoal Technology. pp3-47, AIT, Bankok, Thailand.

Miah M.A., M.A.Baqui, M.D. Huda, and M.Nasiruddin. 1999. Rice Husk Briquette as Alternate Fuel in Bangladesh. Agricultural Mechanization in Asia, Africa and Latin America. Vol. 30 No.2, pp.63-71.

Moral, M. N.A. 2001. Development of Biomass Briquetting System Suitable For Bangladesh. Proc. Nat. Semi. on Utilization of Renewable Energy in Rural Area of Bangladesh. Edited Hossain, Rahaman, Mollah and Alam. Bangladesh Institute of Technology(BIT), Khulna. Asian Institute of Technology (AIT), Bangkok.pp.11

The daily Protmom-Alo 2001. Fourth Population Census from 23 - 27 January 2001.pp..1

Alam 2002. Improved Rice Husk Briquette technology, report submitted to Research and Extension on Farm Power Issue (REFPI).pp.7.

Fig. 1 Photographic view of a traditional RHB machine

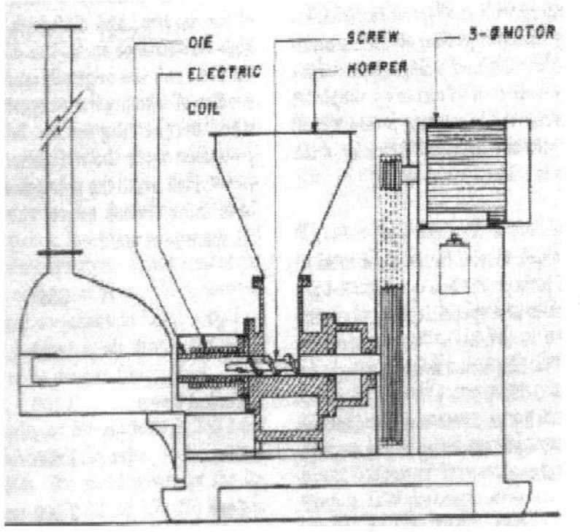

Fig.2 Cross section view of a traditional RHB machine

Fig.3. Photograph of heating coil made by Pipe (Pipe coil)

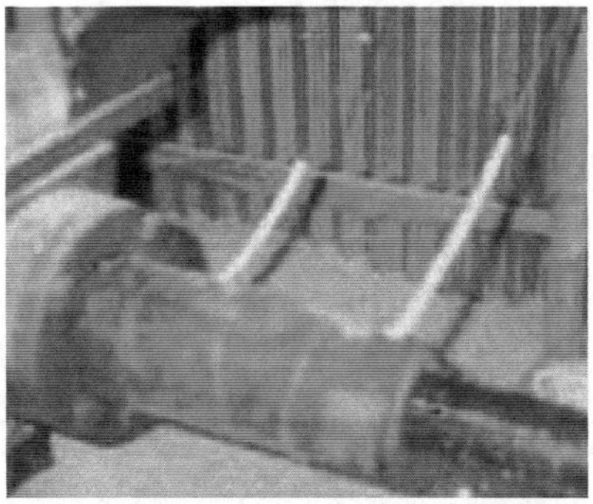

Fig.4 Puti coil die heating system is in operation

Fig.5 Kerosene burner is being used to produce RHB

Fig. 6 A screw (made of mild steel). New screw and B. Wear screw

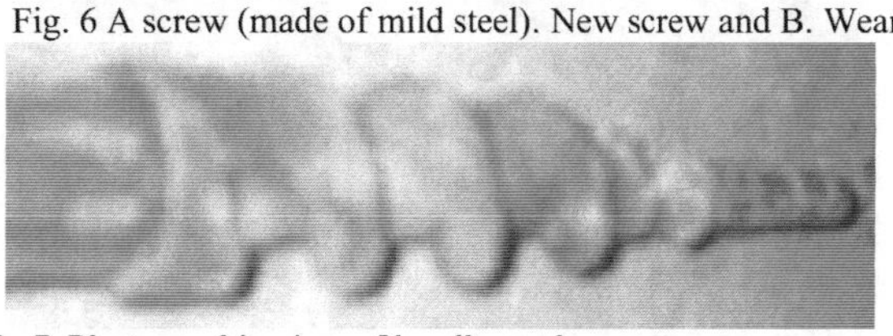

Fig.7. Photographic view of locally made screw.

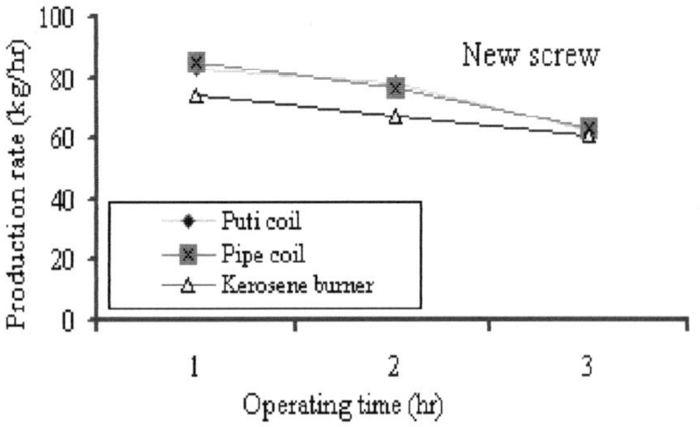

Fig. 8 Production rate of briquette for a new screw at different die heating systems

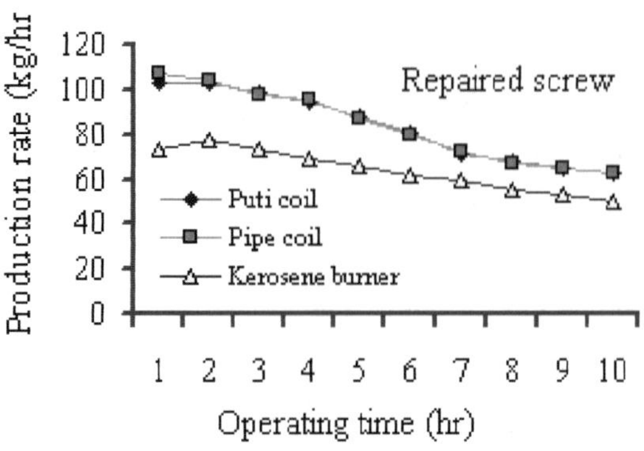

Fig. 9 Production rate of briquette for a repaired screw at different die heating systems.

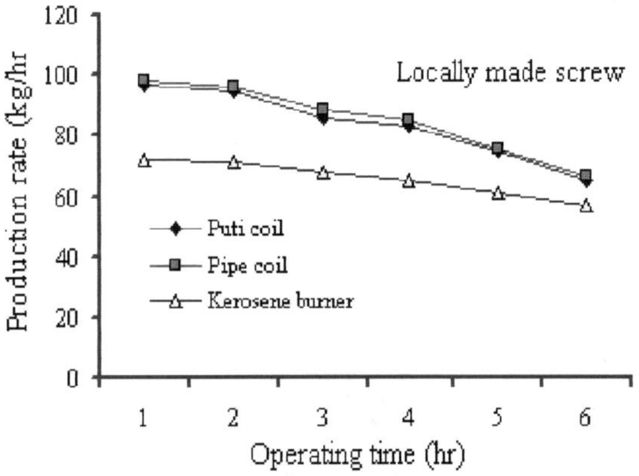

Fig. 10 Production rate of briquette for a locally made screw at different heating systems.

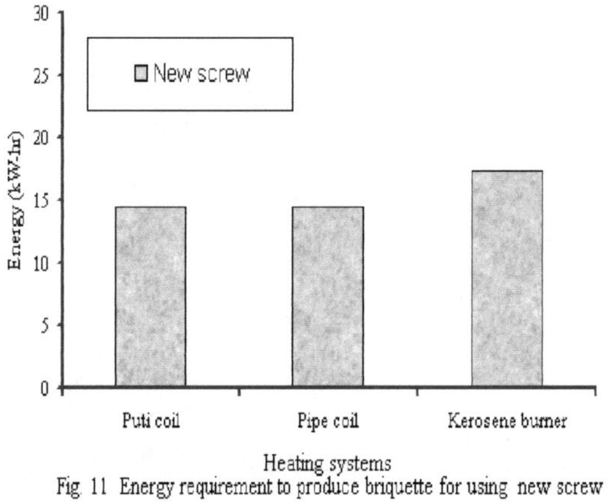

Fig. 11 Energy requirement to produce briquette for using new screw and different die heating systems.

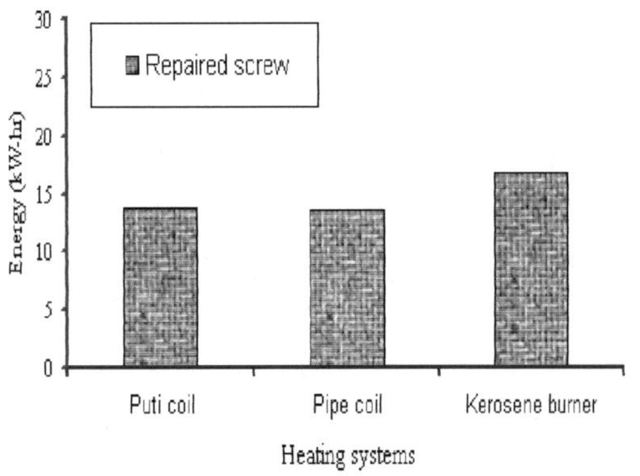

Fig. 12 Energy requirement to produce briquette for using repair screw and different die heating systems.

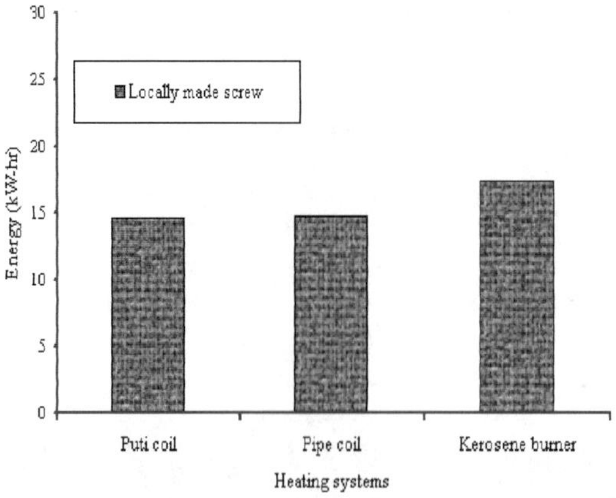

Fig. 13 Energy requirement to produce briquette for using locally made screw and different die heating systems.

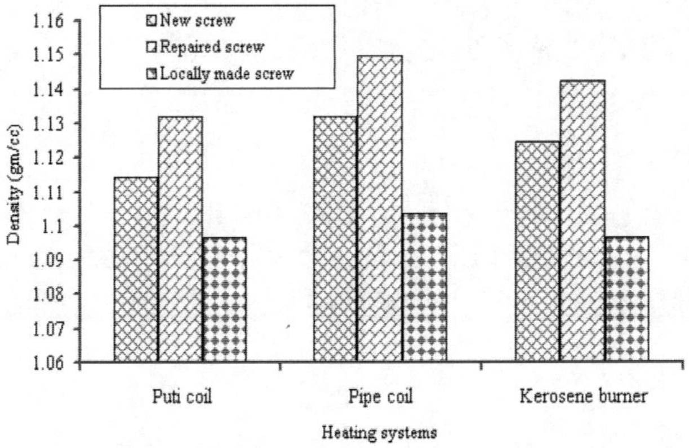

Fig: 14 Variation of density of briquette for different heating systems and screws

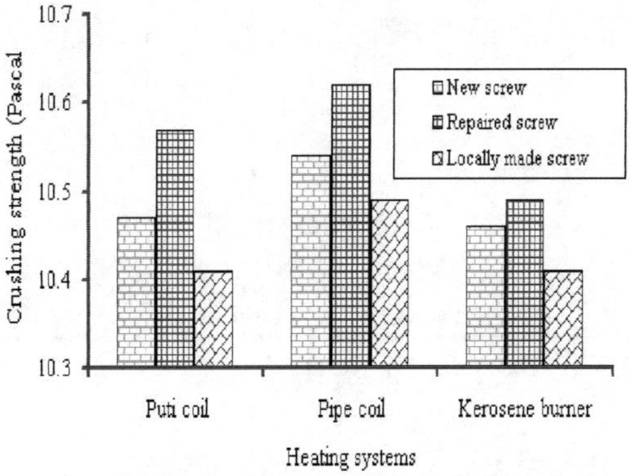

Fig: 15 Variation of crushing strength of briquette for different heating systems and screws.

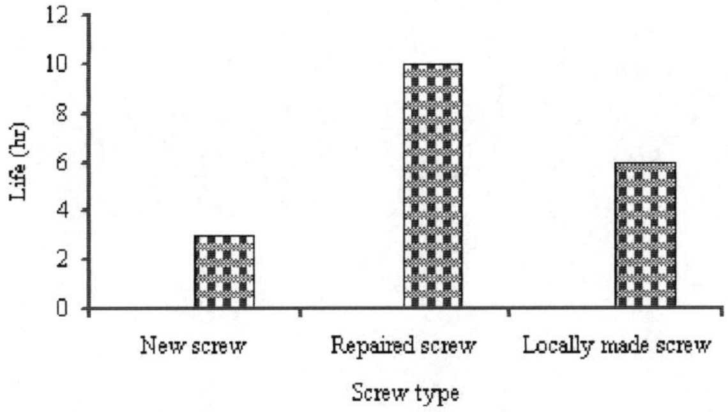

Fig.16 Variation of working life for different scerws

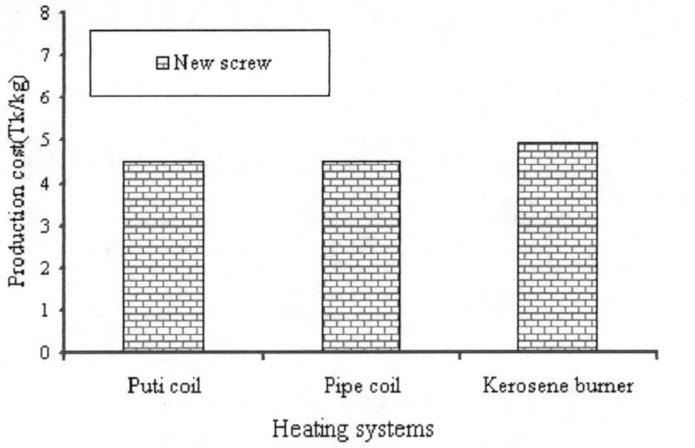

Fig. 17 Production cost for different heating systems when operating with a new screw

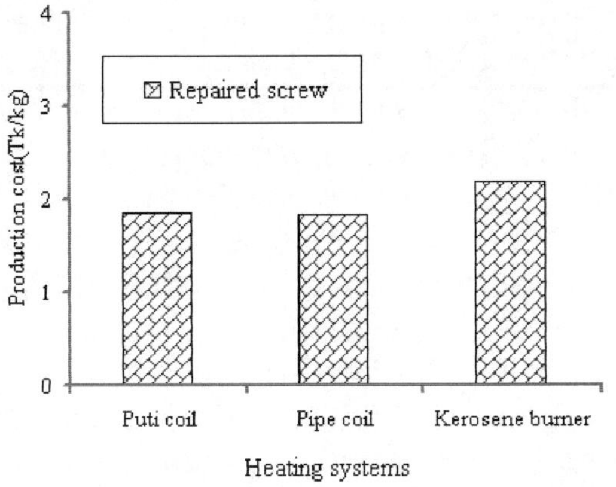

Fig. 18 Production cost for different heating systems when operating with a repaired screw.

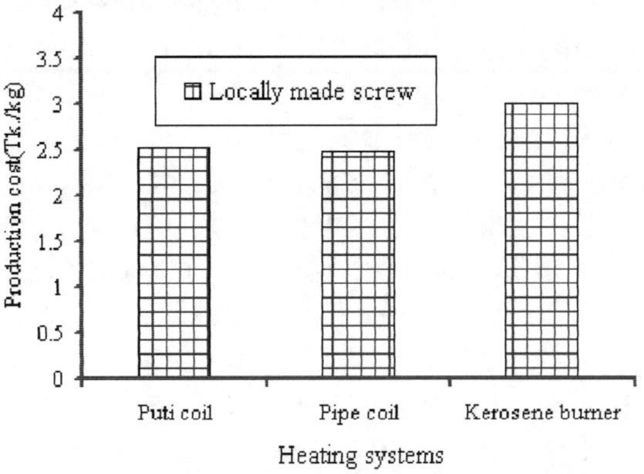

Fig. 19 Production cost for different heating systems when operating with a locally made screw

PRODUCTION OF SYNTHESIS GAS BY BIOMASS GASIFICATION

James T. Cobb, Jr., P.E., Associate Professor Emeritus, University of Pittsburgh
1249 Benedum Hall, Pittsburgh, PA 15261, 412-624-7443, cobb@engr.pitt.edu.

Foreword

As background to the main topic, cofiring biomass to coal gasifiers that are producing synthesis gas, and wood gasifiers that produce synthesis gas, diluted with nitrogen, to inject as fuel into utility boilers will be highlighted. After highlighting these two preliminary topics, the main topic will be presented.

Cofiring Biomass to Coal Gasifiers

Tampa Electric's slurry-fed Texaco gasifier (Polk Power Station 260 MWe Unit #1) has cofired 5% biomass with coal without any major problems (http://www.treepower.org/TECO/polk-cofiring-testburn.pdf).

The Shell gasifier at Nuon's 250 MWe Dernkolec Power Plant at Buggenum, Netherlands has cofired 34% biomass with coal, also without any major problems (http://www.netl.doe.gov/publications/brochures/pdfs/Gasification_Brochure.pdf). The biomass fed to the Buggenum plant has included sewage sludge, chicken litter and demolition wood. The Shell gasifier utilizes a dry feed system.

Firing Biogas to Utility Combustors

In 1994 FERCO installed a 7 MWe SylvaGas unit at the 50-MWe wood-fired McNeil Station in Burlington, VT. The unit was sized at 40 MWt and produced electricity at $0.08/kWh. The gasifier island cost between $11 and $12 million. The gasifier has been shut down because of economics. [This technology is described later in this tutorial.]

In 1997 Foster Wheeler built an air-blown ACFB gasifier in Lahden Lämpövoima Oy's Kymijärvi power plant at Lahti, Finland, having a capital cost of $15 million and an input of 40-70 MWt, depending upon the moisture content of the fuel, which includes residues of fuels, paper and textiles, wood and peat. Lahden Lämpövoima Oy is a Finnish power company producing power and district heat for the city of Lahti. The maximum power capacity is 167 MWe and the maximum district heat production is 240 MW. On an annual basis, the available amount of biofuels and refuse fuels in the Lahti area is enough to substitute about 15 % of the fuels burned in the main boiler, equaling a maximum of 30 % of coal. Fuels like REF (Recycled Fuel), some wood wastes and shredded tires contain different types of solid impurities (nails, screws, metal wires, and concrete). The air distribution grid and the bottom ash extraction system were designed to remove the fuel-borne impurities from the gasifier. The hot product gas is led through the air preheater to two burners, which are located below the coal burners in the boiler.

Status of Commercial Industrial-Scale and Dispersed Biomass Gasifiers

Numerous industrial-scale biomass gasifiers are in operation in Europe for power production. The European Commission recently published a review of biomass gasifiers for liquid fuels production: "Status and Perspectives of Biomass-to-Liquid Fuels in the European Union." Many European biomass gasifiers are derivatives of coal gasifiers.

Numerous industrial-scale MSW gasifiers are in operation in Japan for waste reduction. Many of these were developed by that nations's steel industry.

Numerous dispersed biomass gasifiers are in operation in the United States for agricultural and silvacultural waste utilization. The commercial industrial-scale sector of biomass gasification in the U.S. is entering the demonstration phase.

The biomass gasifiers operating throughout Europe, Japan and the U.S. exhibit a wide range of feed capabilities, gasifier characteristics and product gas cleanup systems. Of particular note are novel feed pretreatment techniques, such as torrefaction, being developed in Europe to improve the flow of feed to biomass gasifiers. Commercial offerers of biomass gasifiers in the U.S. are focused in the demonstration phase on marketing and on further development. Development is directed toward (1) expanding design and operating parameters (feed systems and gasifier characteristics) to utilize a wider variety of biomass feeds; and (2) developing and designing gas cleanup systems to remove tars that interfere with combustion turbines, membrane systems, and catalysts for the production of hydrogen, methanol, ethanol and other chemicals.

The lead USDOE biomass gasification entity, the National Renewable Energy Laboratory (NREL), is working on a thermal platform for biomass conversion. This work is directed toward (1) developing new gasifiers (e.g., the BCT and Startech systems) for conversion of various waste biomass to clean synthesis gas; and (2) providing engineering guidance studies of developed gasifiers (e.g., the SilvaGas and GTI systems) for various uses. The BCT, Startech and SilvaGas systems are described below.

Three Recent Feasibility Studies

The author has recently participated in three feasibility studies:

1. determination of feedstock availability for a proposed wood gasifier in Philadelphia to produce fuel gas for refinery processes (June 2003);
2. identification of a gasifier for a business plan to produce synthesis gas from refuse derived fuel for an electric power project in Pittsburgh (June 2006);
3. identification of a gasifier for a business plan to produce clean synthesis gas from various possible feeds – coal, wood and municipal solid waste – in the southeastern part of the U.S. (August 2006).

Feedstock Availability in Philadelphia

The proposed gasifier was to have been installed in an industrial park to supply synthesis gas to a major petroleum refinery about two miles distant. The 2003 study found sufficient amounts of clean urban wood waste within roughly a circular region with a radius of about five miles around the gasifier. It also found sufficient amounts of refuse derived fuel (RDF) from resource recovery facilities in southeastern Pennsylvania, northern Delaware and southwestern New Jersey. The project has experienced difficulties in being implemented. One impediment has been a concern for increased truck traffic delivering clean wood and RDF through city streets to the gasifier.

Gasifier Selection in Pittsburgh

This 2006 study by Mon Valley Power demonstrated the commercial feasibility of using RDF from new resource recovery facility along the Monongahela River in southern Allegheny County, as the feedstock to a gasifier producing synthesis gas for a standard gas-fired steam boiler to produce 80 MW of electricity. Funds to develop the business plan were provided by the Energy Harvest Program of the Pennsylvania Department of Environmental Resources.

The first step in selecting a gasifier for this project was a broad review of the internet. Detailed reports of two recent surveys were found. Both were conducted by agencies of large metropolises who were seeking guidance on selection of urban waste gasifiers for synthesis gas production. The first survey was carried out in 2004 by the New York City Department of Sanitation (http://www.nyc.gov/html/dsny/downloads/pdf/pubnrpts/swmp-4oct/appendix-f.pdf). Eight MSW gasifiers were found to be of interest. The second survey was carried out in 2005 by the Los Angeles County Department of Public Works (http://ladpw.org/epd/tf/Attachments/SubCommittee%20Attachments/CT_Eval_Rpt_App.pdf). Nine MSW gasifiers were found to be of interest. The gasifiers of interest, found by these two surveys, were then examined in detail, using the descriptions included in the surveys and found on the websites of the companies supplying the gasifiers. Contacts were made with several of the gasifier suppliers and the BIOSYN gasifier (Enerchem) was ultimately selected.

The BIOSYN gasifier is a bubbling fluid bed system, operating at 16 atmospheres and 1832°F. Tars in the hot product gas are cracked in a second chamber, producing a clean synthesis gas at 1,472°F. A 7 MWe demonstration plant, using this technology, was started up in Ribesalbas, Spain, in August 2003. The principal feed to this plant consists of contaminated plastic containers/wrappings, but the process is advertised to be capable of feeding MSW, plastic, wood waste and RDF as well.

The final report containing the business plan for this project was presented in August 2006 to the PADEP. Like the project in Philadelphia, this one has experienced difficulties in being implemented. One impediment has been its inability to negotiate a power purchase agreement.

Gasifier Selection in Southeast U.S.

This second 2006 study demonstrated the commercial feasibility of establishing a privately-financed project to feed 300 T/D of MSW from a nearby landfill, clean forest wood waste from on-site forest land and coal from the current coal broker, to a gasifier producing clean (tar-free) synthesis gas for a standard gas-fired steam boiler to produce 80 MW of electricity.

Gasifiers considered for this project (in descending order of interest for this application at the time of this study) were:

1. Westinghouse Plasma Gasifier (Geoplasma)
2. Plasma Gasification Vitrification (Solena Group)
3. Plasma Converter (Startech Environmental Corporation)
4. Thermoselect Gasification Technology (Interstate Waste Technologies)
5. British Gas Lurgi (Allied Syngas Corporation)
6. Pyromex (Innovative Logistics Solutions)
7. SylvaGas (Biomass Gas and Electric LLC)
8. Bioconversion Technology System (Green Energy Corp)
9. Consutech (BRI)
10. Twin Rec UEP Gasification and Ash Melting Technology (Ebara)

[Brief descriptions of these ten gasifiers are given below.]

At this writing discussions on the gasification process are being held with Westinghouse Plasma Corporation, and the private financer is considering whether to proceed with the project. If the project does go forward, the balance of system would likely be designed by Geoplasma.

Conclusions on the Status of Biomass Gasifiers and the Three Studies

1. Firing biogas to utility boilers is commercial in Europe.
2. Commercial biomass gasifiers are well established in Europe (for power production) and in Japan (for waste reduction).
3. NREL is preparing engineering guidance studies and developing new biomass gasifiers.
4. Commercial biomass gasifier suppliers are adjusting and marketing their systems in the U.S.
5. The economic, environmental and regulatory situation is improving for biomass gasification in the United States.
6. Ample biomass is present in urban areas for several industrial biomass gasifiers per community.
7. The BIOSYN gasifier (Enerkem) has been selected for a project concept in Pittsburgh for power generation.
8. The Westinghouse Plasma Gasifier (Geoplasma) has been selected for a project in southeastern U.S. for cogeneration of steam and clean synthesis gas.
9. Nine other biomass gasifiers have been identified as of interest to the developer of the southeastern U.S. project and have been described for the project developer.

Brief Descriptions of Ten Gasifiers

Westinghouse Plasma Gasifier

The Westinghouse Plasma Gasifier is essentially a classic down-flow moving bed system, operating at one atmosphere, but at an elevated temperature of 2,282°F. The thermal energy required to reach this high temperature is provided by Westinghouse Plasma torches, located in the bottom of the gasifier and firing into a bed of carbon. The inorganic material from the feed is melted in the carbon bed, forming glass aggregate and metal nodules that emerge from the bottom of the unit. Westinghouse Plasma has designed a donut-shaped chamber, above the moving bed, where tars and other hard-to-gasify molecules reside for 0.5 to 1.0 minutes and are cracked. Carbon conversion is 100%. Any material may be gasified or melted – simply placed without preparation into the top of the gasifier. See www.westinghouse-plasma.com.

An eight-MWe commercial plant, feeding 300 T/D of MSW, was started up in April, 2003 in Utashinai, Japan, having a capital cost of $65,000,000. For a plant feeding 480 T/D of MSW (23 wt% H_2O, 23% ash, HHV = 5,660) and 20 T/D of coke, the product slate (N_2-free) is 40.37 mole% CO, 15.88% H_2, 37.33% H_2O and 3.55% CO_2 with a HHV of 276 Btu/ft^3. A 3,000 T/D commercial plant, having a capital cost of $425,000,000, has been announced for startup in 2008 in St. Lucie County, Florida. The feed to this plant will be MSW and water treatment plant sludge, and the plant will be comprised of six 500 T/D gasifier modules. Steam will be sold to Tropicana Products for conversion to electricity in Tropicana's nearby existing steam turbines.

Plasma Gasification Vitrification (Solena Group)

Like the Westinghouse Plasma Gasifier, the Solena gasifer locates plasma torches in the bottom of the gasifier to vitrify inorganics in the feed, forming glass aggregate. Less energy is injected through the torches than in the Westinghouse system, and a carbon-based catalyst used to enhance gasification in the bed above the torches. Acid gases, volatile metals and particulate matter are removed from the cooled gasifier effluent, producing a gas sufficiently clean to be fed to gas turbines and fuel cells. See http://www.solenagroup.com/html/home.asp.

Solena Group is developing 1-MWe Integrated Plasma Gasification/Fuel Cell project to process 1-5 T/Hr of solid waste. It is planning a 10-MWe plant in Malaysia, using padi husks, that will cost $250 million. It is also planning a number of 130-MWe Integrated Plasma Gasification Combined Cycle plants in the eastern U.S. to use waste coal and coal fines. Solena Group's principal partner is Stone and Webster.

Plasma Converter (Startech Environmental Corporation)

Shredded feed is injected to this gasifier using a pump, screw or ram, depending upon the consistency of the feed. The plasma torch is located in the top of the gasifier and is can be directed to dissociate organics and to melt inorganics in the feed, forming clean gas and glass aggregate. Acid gases, volatile metals and particulate matter are removed from the cooled gasifier effluent. The product gas, after cleaning, can be fed directly to gas turbines and fuel cells. If hydrogen is the desired product, it can be separated from the other gas components using the Starcell™ technology. See http://www.startech.net/.

Startech Environmental Corporation has formed a strategic alliance with Hydro-Chem, a division of Linde. Two plants are planned, one being a 200 T/D MSW plant in Panama and the other being a 10 T/D PCB/POP plant in China. Numerous other plants are under negotiation in the Far East.

Thermoselect Gasification Technology (Interstate Waste Technologies)

There are four steps in this process – MSW compaction (which degasses the feed), pyrolyzation at 570°F (which produces volatile hydrocarbons and char), oxygen-blown entrained char gasification at 2,200°F (which produces clean, dioxin-free synthesis gas) and shock-cooling to 200°F. Inorganics are heated to 3,600°F, converting them to inert material. The synthesis gas can be used for power or chemicals production. See http://www.interstatewastetechnologies.com.

A 606 T/D plant began operation in Feb 2005 in Kurashiki, Japan, and four other Japanese plants have been started up since then. The stated cost of this technology is $150K/daily ton of MSW fed ($5.5 million/MW).

British Gas Lurgi (Allied Syngas Corporation)

This gasifier consists of a Lurgi moving bed, below which is placed a British Gas slagging bottom. The feed is injected through a lock hopper. The gasifier operates at 1450°C and 25 bar. It is blown with steam and oxygen, and produces clean, nitrogen-free synthesis gas, even when MSW is the feed. The synthesis gas can be used directly, for example, to produce methanol. See http://www.kiche.or.kr/files/2006SS8.pdf.

An 840 T/D BGL slag-bed gasifier has been operating since 2003 at SVZ Sustec Verwertungs-Zentrum (recycling center) in Germany. See http://www.gasification.org/DOCs/2005_Papers/23SHEL.pdf and http://www.advantica.biz/PDF/gasifier_presentation_barca.pdf \. Allied Syngas Corporation, Wayne, PA, is the U.S. licensee of this technology.

The BG slagging bottom was developed originally in Westfield, England. Global Energy Inc., which owns Wabash River Energy Ltd., has purchased the Westfield facility and is converting the developmental unit there to gasify renewable fuels. See http://www.globalenergyinc.com/.

Pyromex (Innovative Logistics Solutions)

Induction (non-contact) heating to 3,000°F provides the thermal energy to this rotary kiln pyrolyzer, constructed from a high temperature alloy. Inorganics in the feed are converted to sand or basalt-like residue. The product, “Pyro-gas,” contains 70% H_2 and CO and no tar. The raw product can be reformed and purified to 99.999% H_2. See http://www.ils-partners.com.

A 25 T/D sludge-fed plant is in operation in Germany. ILS states that it can design the pyrolyzer to handle 500 T/D MSW.

SylvaGas (Biomass Gas and Electric LLC)

The SylvaGas system consists of coupled circulating fluid-beds. One of the beds operates at 1800°F, and combusts char from the other bed and simultaneously heats sand to be circulated to the other bed. That other bed operates at 1500°F, and pyrolyzes biomass, using heat from sand that is circulated from the first bed. The gas produced by the pyrolyzer can be cleaned for use in fuel cells or in chemicals production. See http://www.biggreenenergy.com/.

A 20 MW plant, converting 400 T/D of clean biomass to synthesis gas, is under construction in Forsythe, Georgia. Its estimated cost is $40 million. A 38 MW plant (750 T/D) is planned for Tallahassee, FL, to also provide synthetic gas to a natural gas distribution system. This plant is estimated to cost $85 million.

Bioconversion Technology System (Green Energy Corp)

In this gasifier hot steam is used to pyrolyze and reform the carbonaceous feedstock. The syngas produced in the process is catalytically converted to ethanol. The system can generate up to 20,000 GPD ethanol from 400 wet (200 dry) T/D of feed material. See http://www.greenenergycorp.com/index.htm and http://www.biggreenenergy.com/.

The process is offered by Biomass Gas and Electric LLC which has an Equipment Supply and Licensing Agreement with Infinity Energy Solutions.

Consutech Gasifier (BRI Energy LLC)

The first stage of this process is a standard step-grate combustor (frequently used as an MSW incinerator) that operates as a gasifier at 950°F using oxygen-enriched air. The second stage is a heat treater that operates at 2000-2250°F and uses minimal oxygen to crack tars. A water scrubber, operating at 97°F then prepares the product syngas for a biological converter that produces ethanol. See http://www.magic-region.com/BRI_Energy_Backgrounder.pdf.

A 1.5 T/D pilot plant is in operation in Fayetteville, AK.

Twin Rec UEP Gasification and Ash Melting Technology (Ebara)

The current Ebara system is an air-blown fluid-bed gasifier that operates at 550°C. Metal-containing char is discharged from the bottom of the gasifier, and the fly ash-containing gas, exiting the top of the gasifier, is fed to slagging combustor that operates at 1400°C. The unit currently is designed for power production only, but Ebara is developing an oxygen-blown system to generate clean synthesis gas for the production of fuels and chemicals. See http://www.ebara.ch/downloads/EBARA%20Manuscript%20Sheffield%20IChemE%202003.pdf .

An 80 MW_t air-blown plant began operation in Feb 2000 in Aomori, Japan. This plant contains two 40 MW_t trains to which are fed shredded residue and sewage sludge. The cost of this plant was $23 million/MW.

Final Conclusions

Biomass can be cofired with coal directly to utility boilerplants and gasifiers.

Biomass can be gasified to produce tar-containing synthesis gas to cofire into utility boilerplants.

Eleven commercial and developmental industrial-scale biomass gasifiers can produce clean synthesis gas for gas turbines and catalytic processes.

Numerous commercial projects utilize these clean biomass gasifiers in Europe, Japan and the U.S.

Improving the Design and Operation of Combined Heat and Power Systems in Refining Plants

Oscar Aguilar, Shell Global Solutions (US) Inc., 3333 Hwy.6 South, Houston, TX 77082

Most refining and petrochemical sites employ combined heat and power (CHP) plants to meet their demands of electricity, shaft power and steam. In this context, CHP systems typically comprise many energy-conversion units (e.g. boilers, steam and gas turbines) that entail substantial investment and large operating costs. Consequently, these plants offer an excellent opportunity to achieve significant savings, but, at the same time, minimising such expenditure also represents a challenging task due to the large number of design choices and complex operating interactions among the equipment.

Shell Global Solutions (GS) through its Utilities Group has provided consultancy to many sites worldwide in order to implement operating policies and/or infrastructure projects that effectively reduce costs and improve the reliability of their CHP plants. Throughout its long experience, Shell GS has adopted and developed several simulation and optimisation tools to understand the degrees of freedom offered by a CHP system in a comprehensive approach.

This paper examines a case study representing a refinery CHP plant subject to variable requirements and constraints (e.g. demands, fuel availability, electricity prices). This system needs to be revamped to reduce operational costs, increase efficiency, improve its reliability, and meet tougher emissions limits. Such objectives, together with the varying parameters, are simultaneously taken into account in order to arrive at a solution that includes, not only the new design, but also a suggested operational policy for the scenarios considered. Finally, a sensitivity analysis evaluates the impact of different situations and constraints on the conclusions for the project.

Keywords: Utilities, energy, optimisation, cogeneration, flexibility.

PORT ARTHUR STEAM ENERGY: BRINGING A MAJOR ENERGY RECOVERY PLANT BACK TO LIFE

Tommy John, *Energy Advisory Group;* **John Gray**, *Energy Advisory Group;* **Ray Deyoe**, *Integral Power, LLC; and* **Paul Gross**, *Liquid Process Technologies, Inc.*

Abstract

Refurbishment of the Port Arthur Steam Energy facility began in early 2005 after key commercial agreements were concluded. The plant, which had been idle since October 2000, was originally constructed in 1983 and 1984 to recover energy from three petroleum coke calcining kilns at the Great Lakes Carbon LLC facility. Major repairs were needed because of extensive damage from sulfuric acid corrosion of the HRSG system and deterioration of water treatment facilities. In addition, major improvements were made including an acoustic cleaning system, multiclones for particulate emission reduction, magnesiun oxide injection for corrosion control, a complete new control system with all new field instrumentation, stack improvements to increase dispersion, and improvements to the HRSG system and water treatment system to improve reliability and reduce maintenance.

Rising energy prices dictated a fast-paced schedule. Following a major reconstruction effort with a peak force of 435 people, the facility was in full operation by August 2005 (less than nine months from commencement) producing approximately 450,000 lb/hr of high pressure steam, the majority of which is sold to the neighboring Valero Port Arthur refinery, and producing 4 to 5 MW of power. By capturing 1800-2000°F heat that would other wise be wasted, the project recovers nearly 5 trillion BTU/year, off setting over 200 tons/yr of NO_x and over 280,000 tons/yr of carbon dioxide that would otherwise be emitted by natural gas combustion.

Some refineries have coke calcining facilities on site, others, such as this facility, are nearby. This project illustrates how the operation of these facilities can be integrated to reduce emissions and save energy. The success of the project can be attributed to management of the project which included innovative inspection techniques, development of the scope of work, design of improvements, and extensive construction and repairs.

Background

The plant was developed, designed, and constructed by Power Systems Engineering in the early 1980's to recover energy from the exhaust of petroleum coke calcining kilns in Port Arthur, Texas. The project included a 2.5 mile steam line to the nearby Gulf Oil refinery (now owned by Valero) and a clarified water line from the refinery. The plant shut down in October, 2000 following expiration of the steam sale agreement. At that time, the plant was owned by Dynegy, which was unable to negotiate a renewal with Premcor (Clark Oil), the then current owner of the refinery.

The facility is located inside of the Great Lakes Carbon (GLC) Port Arthur Plant, a world scale operation for calcining petroleum coke primarily for use in electrodes for aluminum processing. The coke calcining process uses rotary kilns to heat petroleum coke to drive out volatile hydrocarbons leaving pure carbon. Most of the heat is supplied by consuming about 15% of the coke feed, but some natural gas is used for temperature control. The exhaust gas from the process is at 1800 to 2000F, and contains sulfur dioxide and particulate matter. Following the kiln is a pyroscrubber, a large refractory lined chamber with residence time and air addition to combust coke particulate matter, VOC, and CO and allow any remaining particulate matter to settle. The pyroscrubber converts about 3% of the sulfur dioxide (SO_2) to sulfur trioxide (SO_3), which can condense as very corrosive sulfuric acid. Absent the heat recovery system, the pyroscrubber exhaust exits through a large diameter refractory lined stack that is 150 feet tall and provides natural draft for the system so that the kilns operate with negative pressure.

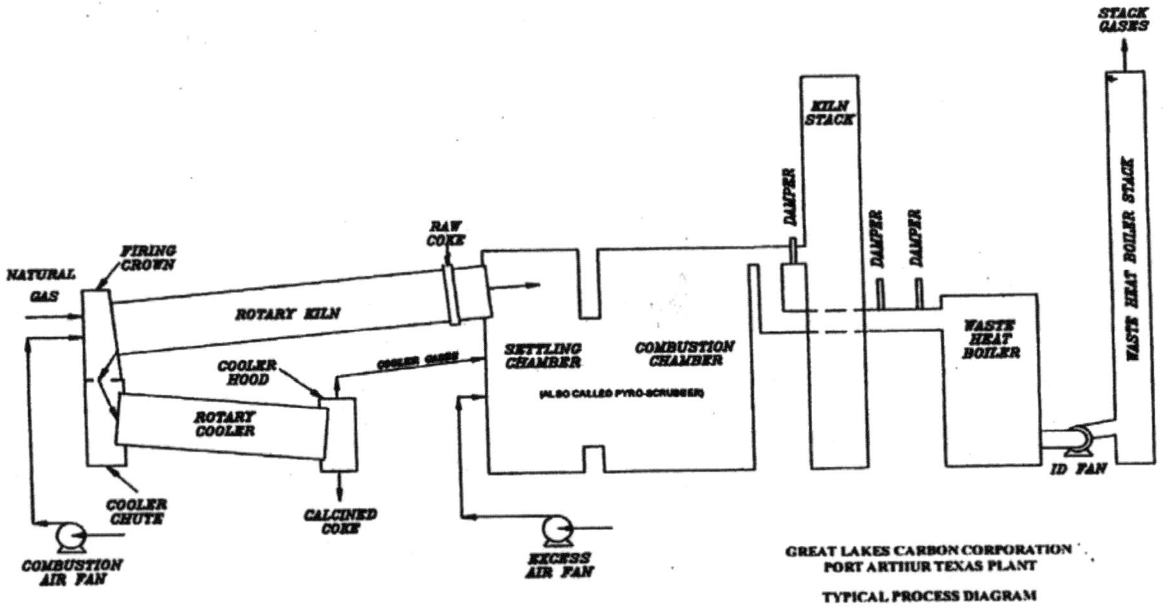

Figure 1- Coke Calcining Process with Heat Recovery

The GLC Port Arthur facility has four kilns of varying size: #5 is the largest and newest, #3 and #4 are identical in size, and #2 is the smallest. Heat recovery systems were installed on Kilns #3, 4, and 5. The heat recovery systems are installed as near as possible to the kilns. The balance of plant is located adjacent to GLC and operated as a stand alone facility.

Heat Recovery System

The Heat Recovery System is designed to produce a single pressure level of superheated steam at 875 psig and 770F. It includes breeching at the bottom of the refractory lined stack (hot stack) with hydraulically operated guillotine dampers; one to isolate the stack, and two to isolate the HRSG so that the kilns can operate with the HRSG out of service. Because of space limitation, the stack isolation damper for #3 is at the top of the stack instead of at the breeching. It consists of four counter weighted quadrants opened by hydraulic cylinders. There is a rectangular, refractory lined inlet duct connected to a membrane wall radiant section. There are two sections of top-supported bare tube superheaters with refractory lined casing and a desuperheater for outlet temperature control. The two top supported bare tube convection sections have 24 inch top and bottom headers with internally lined casing; the first section is refractory lined and the second has insulation with a stainless steel liner. The horizontal finned tube economizer originally had casing with internal insulation and a steel liner. After less than 10 years of operation, the casing was replaced with an externally insulated system because of severe sulfuric acid dew point corrosion. There are a total of 20 Diamond Power rotary steam soot blowers in each HRSG covering the superheater through the economizer. The carbon steel outlet duct has external insulation. At the outlet of the economizer is the Induced Draft (ID) fan with a DC variable speed drive. The exhaust stack (cold stack) had an internal refractory lining with an emission test platform at which there were continuous opacity monitors (COM).

Balance of plant

A 10-inch clarified water pipeline delivers water from the refinery, which is primarily used, after treatment, as boiler makeup water. Makeup is 100% because there is no condensate return. The boiler makeup water system has the capacity of 1200 gpm, and consists of three parallel strong acid cation and strong base anion ion exchange trains followed by two parallel mixed bed polishers. Regeneration is with sulfuric acid and caustic. Regeneration waste is collected and neutralized in a 136,000 gallon tank. There are two 500,000 gallon demineralized water storage tanks. Two deaerators with a common storage tank operate at approximately 63 psig to supply feed water at about 310F to avoid sulfuric acid dew point corrosion in the economizers. Two boiler feed water pumps have DC variable speed drives similar to the ID fans, and a third has a backpressure steam turbine drive. A 6.5 MW back pressure turbine generator produces power for the coke calcining and heat recovery facility and exhaust steam for the DA heaters. A substation provides interconnection to the Gulf States Entergy 69 kV grid. The original control system was a Rosemount Diogenes system located in a central control room.

Condition of the Facility

When the steam contract with the Premcor refinery expired, Dynegy ceased operation, and eventually turned over the facility to GLC. Minimal steps were taken to preserve or mothball the facility. Feed water with lay-up chemicals was pumped into the HRSG under a nitrogen blanket, but was drained after the nitrogen leaked off. Despite this, the water side was generally in good condition because a good magnetite layer had been established by the treatment program. The air side of the HRSG was left open and full of corrosive coke dust. Water was drained from the demineralizer plant, but the resin was left in the vessels. The turbine generator and other turbo machinery was shut down with the rotors still in the case. The entire plant staff was laid off. The admin building housing offices, documents, control room, steam turbine/generator, BFW pumps and maintenance facilities were eventually abandoned under lock and key.

Return to Operation

Integral Power LLC (IP) recognized the opportunity and began development activities in July 2002. The key step was reaching a new steam agreement with Premcor in late 2004. In order to fund the refurbishment of the plant, IP arranged for financing with American Industrial Partners (AIP), a private equity firm that invests in established industries. AIP had a substantial stake in GLC, and recognized that the project would have an excellent return on investment plus add value to GLC. With AIP backing, IP initiated inspections, and developed a preliminary scope of work and a preliminary budget and economics. The economic analysis indicated a strong rate of return with relatively low risk, so a decision was made to proceed on the project on a fast track without a detailed and accurate cost estimate. AIP and IP formed Port Arthur Steam Energy (PASE) and contracted with Energy Advisory Group, Inc. (EAG) to manage the reconstruction effort.

Environmental

The coke calcining operation is inherently a major source for particulate and sulfur dioxide emissions, so environmental management is a major focus. GLC holds the permit for air emissions from the kilns through both the hot stack and cold stacks which includes limits for particulates, sulfur dioxide, and opacity. Cooling the kiln exhaust to recover energy could, under certain conditions, cause sulfuric acid mist to form creating visible particulate emissions. PASE developed a control strategy of injecting magnesium oxide to react with sulfur trioxide to form solid magnesium sulfate which would be removed with multiclones. The multiclones have the additional benefit of reducing other particulate emissions. A simple, portable system for magnesium oxide injection was selected for use until the optimum injection point could be determined.

Dispersion modeling predicted excessive sulfur dioxide ground level concentration under certain conditions. This is prevented by increasing the height of the cold stacks to 185 feet from 150 feet (175 feet for #5). Continuous Opacity Monitors (COM) are required by permit on the cold stacks. Since the only new equipment was for emission reduction, the project qualified for a Standard Permit from the Texas Commission for Environmental Quality (TCEQ), which allowed the project to be completed on an expedited basis. A wastewater discharge permit was obtained by PASE, primarily for neutralized waste water from the demineralizer. Ash collected from the multiclones is considered non-hazardous and is disposed of at a nearby landfill.

Inspection, Repairs, and Improvements

Although the project commenced based on preliminary inspection, scope, and budget, a detailed inspection of the entire plant began immediately to better define the overall scope of work. This activity was carefully coordinated with other activities, and the scope of work was updated as required.

A visual inspection of the inlet duct showed that the firebrick lining in #5 needed replacement. The old refractory was removed by robot, new anchors installed, and plastic refractory rammed in place. Refractory repairs were made in #3 and #4.

Throughout the life of the plant, there were problems with acid dew point corrosion on the outside of the membrane tubes in the radiant section. The membrane wall is open at the corners and covered with insulation and gas tight casing, so the outside of the wall is exposed to flue gas. When the unit is down, and the metal cools below the dew point, significant damage occurs in a very irregular and unpredictable pattern. The only way to inspect for damage was to remove all casing and insulation from the radiant sections and to hydro test the tubes. A few failures could be seen by visual inspection, but most were detected by multiple hydro tests. About 10% of the membrane wall was replaced in #5, a lesser amount in #4, and none in #3.

Visual and ultrasonic (UT) inspection of all sections of the HRSG was done. The superheaters had catastrophic external corrosion of tubes and headers at the lower end from coke dust. A refractory lined tube sheet with sleeves for the tube to pass through protects the headers from the gas path. Accumulation of coke dust on the tube sheet and inside the sleeves badly corroded the tubes during the extended outage. The entire superheater coil was replaced. When the casing was removed to inspect the top headers, it was discovered that the top supports had failed. Accumulation of coke dust below the bottom headers prevented downward thermal growth, which caused the guides that held the top headers to break away from the structure. A similar accumulation of coke dust in the convection sections caused the bottom casing to split open. For a solution, PASE expanded space and access below the superheater and convection sections to allow removal of coke dust to prevent damage from build up.

Inspection of convection sections showed external corrosion of tubes at the bottom from contact with coke dust and at the top from contact with failed ceiling refractory. Internal Rotary Inspection System (IRIS) inspection of convection sections was used to detect external corrosion of tubes and erosion by the rotary sootblowers, which was more prevalent at the opposite end from the sootblower head, probably due to difference in thermal expansion between the lance and the coil. About 10% of convection section tubes were replaced. All refractory was in bad condition from sulfur attack from the outlet of the radiant section through the first convection section and was replaced. The liner in the #2 Convection Sections was in relatively good condition and was repaired without removing tubes for access. Visual and UT inspection of the boiler casing showed many holes and thin spots, so it was all replaced

Inspection of the economizer indicated that it was overall in good condition; there was some evidence of flow accelerated corrosion near some return bends, but it was not severe enough to warrant tube replacement. Visual inspection of the fan rotor showed some corrosion around shaft seals and erosion/corrosion of fan blades. Internal (visual) and external (UT) inspection of stacks revealed that the refractory liner had failed in lower sections, and significant metal loss, particularly in the upper sections. The refractory lining and existing stacks were removed down to 55 feet (130 feet for #5) above grade and new, extended stacks with dynamic balancer to control vortex shedding vibration were installed. The stacks are unlined because experience shows that internal insulation promotes dew point corrosion because the metal temperature is lower.

An acoustic cleaning system was supplied by Advanced Acoustical Technologies. Two horns in each unit generate sound in the 40 to 50 hz range tuned to create acoustical resonance within the HRSG gas path. The location and frequency of the horns was determined by modeling of the system. The high amplitude sound waves penetrate the boundary layer around the tubes causing ash collected on the tubes to be re-entrained in the gas flow. A blower provides a common air supply for intermittent operation of each horn.

Visual, UT, and infrared thermographic inspection of stack cap on #3 showed local corrosion from acid dew point near attachments such as emission test nozzles and stiffeners. The outward projection acts as a cooling fin to conduct heat away from the stack shell causing a local temperature below the dew point. Visual inspection indicated that the segments needed replacement, because of dew point corrosion of the carbon steel backing. The entire stack cap was replaced and the hydraulic system refurbished.

Infrared camera inspection of electrical components was performed, and repairs made. Inspection of demineralizer vessels revealed that the rubber lining was badly deteriorated and unfit for service, so it was removed and replaced with natural rubber. Spark testing of demineralized water tanks revealed that the bottom 6 feet of the lining needed to be repaired. Replacement included the damaged sections of the under drain system in several of the DM vessels, all diaphragm actuated valves, acid and caustic regeneration skids, and conductivity, pH and silica monitors. Hydro testing of the clarified water line showed that the FRP pipe was deteriorated and was replaced with high density polyethylene pipe with fused joints.

The turbine generator and BFW pump steam turbine drive were disassembled, inspected, and refurbished; all pumps were sent to repair shops to be refurbished (several pumps required replacement); and the air compressors were replaced with Atlas Copco dry screw and the instrument air dryer replaced. The instrumentation scope of work included a Emerson Delta V Distributed Control System with Rosemount Smart transmitters, refurbish or replace all control valves, new opacity monitors, and voice and digital communication link between PASE and GLC control rooms. The electrical scope of work included refurbishing switchgear (replacing several breakers) and adding switchgear for new components; replacing DC motors for ID fans & BFW pumps because the original motors were obsolete and parts are no longer available; replacing uninterruptible power system, and checking all wiring. All AC motors were tested, dipped and baked, and retested; or were replaced. The generator was dried out and tested, and the generator controls and the SCR for DC drives refurbished.

Project Management

Successful implementation of this project was the result of the combined efforts of a strong and effective team that included the project developer, Integral Power, AIP who provided financing, and EAG, who provided the project managers. PASE was its own Project Manager, and all project management activities were performed on site. Several former employees of the plant were brought in at the beginning to provide a core for operations and maintenance including safety supervision. GLC assigned a full time project manager to coordinate activities and there were frequent meetings with Valero to coordinate refurbishment and operation of steam and water lines. There was a strong emphasis on safety throughout the project. PASE Operations was onsite from the very beginning of the project with the responsibility to develop and implement a safety and security program. Each contractor was responsible for having a safety program and coordinating with other contractors.

A local industrial contractor provided support and much of the balance of plant work, but the majority of the field work was performed by specialty contractors including boiler, refractory, instrumentation and electrical, high voltage electrical, inspection, turbomachinery, rubber lining, scaffolding, and insulation. Major equipment including the multiclones, Distributed Control System, acoustic cleaning system, demineralizer regeneration skids, air compressors, and stacks were designed and fabricated by suppliers. A local engineering contractor supplied engineering services as needed. Local shops refurbished pumps, safety and control valves, and motors.

All onsite work was performed on a time and material basis because of limited time, the need for a flexible scope, and to promote coordination and team work among contractors. All contractors were required to keep schedule information up to date for integration into the master Primavera critical path schedule, which was up dated continuously and reissued weekly. As the project developed, there were significant scope changes as the result of inspection and opportunities for improvements. The budget was updated continuously. The project team practiced very disciplined spending, and despite the scope changes and additional cost for the fast track schedule, the overall cost was within 30% of the preliminary budget.

The first quarter of 2005 was devoted primarily to inspection, scoping, and contractor selection. The second quarter and the first half of the third quarter of 2005 was a period of intense construction with manpower peaking at about 435. The worksite was extremely congested because the HRSG units were an add-on installed as near as possible to the kilns. The coke calcining facility remained in full operation throughout the construction period. Any work requiring a kiln outage was scheduled for a period corresponding to kiln maintenance.

Startup

The DM water system was started up on July 23, 2005. Following commissioning of the Valero steam system, PASE began steam delivery on August 9, 2005. Overall the startup went well. There were some electrical problems with generator controls that were eventually traced down and rectified. The magnesium oxide for sulfuric acid control was originally purchased as magnesium hydroxide (converts to magnesium oxide when heated) slurry, which was subject to plugging. A re-formulated product that remains in suspension cured the plugging problem. A technique was developed for measuring the pH of the fly ash collected in the multiclones and adjusting the Magnesium Oxide feed rate to produce a neutral pH. A portable SO3 dew point monitor was purchased and is used to periodically check the system performance. The fly ash is transferred by conveyor into super sacks which are hauled away as non-hazardous waste for landfill. Initially there were a few relatively minor conveyor mechanical problems that were fixed. In order to cure refractory, the HRSGs were started up very slowly. The boiler feed water makeup was cold, and it took several days to get the temperature of the boiler feed water in the storage tank to the normal operating temperature of 310F. Acid condensation was observed in the exhaust end of the HRSG, but it stopped when the design feed water temperature and steady state stack gas temperature was reached.

Performance

The PASE Heat Recovery plant availability is an impressive 99.5%, which is the result of good operations and maintenance practices, the efficacy of the reconstruction, and close coordination of maintenance activities with GLC kiln outages. Since the kiln availability is over 92%, the plant is a very reliable source of steam to the Valero 600 psig steam system consistently producing over 500 MMBtu/hr. Total energy recovered is nearly 5 trillion BTU/year, off setting over 200 tons/yr of NO_x and over 280,000 tons/yr of carbon dioxide that would otherwise be emitted by natural gas combustion.

About a month after startup, Hurricane Rita struck the Texas-Louisiana coastline a few miles east of the plant site. There was no major physical damage to plant, but it was down 20 days because of loss of power. Although the steam turbine generator can operate the complex isolated from the grid, there was no black start capability. Subsequently PASE installed a 375 kW diesel generator for control and auxiliary power for startup and GLC installed a 2MW gas fired generator that can provide enough power to start at least one kiln and heat recovery system to provide steam for the steam turbine generator. This allows the entire operation to come up and operate independent of the grid.

Testing showed that the performance of the magnesium hydroxide system could be improved by better distribution of the reagent in the gas stream. The simple system with a single injection point in the inlet duct was replaced with a new nozzle design and six nozzles per boiler duct. The system reduces SO3 by 80 to 85% when using about 50% excess MgOH which reduces the acid dew point from above 310F to 280-290F. Internal inspection of the HRSG air side shows that the MgO is providing good corrosion control and that the acoustic cleaning system is effective in keeping the heat transfer surface clean. A couple of finned tubes were installed on the outside rows of the #2 Convection Section to test for fouling, and the acoustic cleaning system keeps them clean.

The COM indicates that opacity is within limits except for short term episodes that occurred an average of once per month. Evidence indicates that the episodes were caused by kiln upsets that reduce the temperature in the pyro scrubber and result in reduced combustion of the coke particulate matter. GLC is making improvements to the temperature control system to fix the problem.

Most of the outages were for repairs to the outside of the radiant section membrane walls in #4 caused by sulfuric acid corrosion. The corrosion is ubiquitous and random, and there is no explanation for why it affects the HRSG units in certain areas more than others.

Conclusions

- Recovery of steam from the exhaust of coke calcining kilns is an effective way to reduce refinery energy consumption and reduce emissions.
- Effective project management will maximize the benefit from a refurbishment project by balancing the benefits of schedule, cost, and reliability.
- An acoustic cleaning system designed to use resonance to enhance the sound energy is an effective way to keep the heat transfer surfaces clean.
- Magnesium Oxide is effective in reducing sulfuric acid dew point corrosion and opacity.

Energy Targeting under Varying Driving Force Distribution Conditions in Heat Recovery Systems

Mahmoud Bahy Noureldin[1], Ahmed Saleh Aseeri[1], and Saleh Al-Hashimi[2]. (1) Consulting Services Department, Saudi Aramco, Dhahran, Saudi Arabia, (2) Chemical engineering department, The Petroleum Institute, Abu Dhabi

Oil and gas supplies are very tight and its markets are suffering high level of uncertainty. The demand for more fossil fuel is increasing and there is no sign it will decline in the foreseen future. Almost all production facilities are doing their best to save as much as possible in their energy bill to keep its competitive edge. Energy conservation and energy efficiency optimization practices have become a normal operation mode for industrial facilities. Optimal driving force distribution in energy systems comes from the combined effect of system's process conditions and hot and cold streams minimum approach temperatures. Optimal driving force distribution in energy systems can have huge effect on energy consumption, utility selection and utility systems as well as energy recovery systems capital investment.

Process changes can result in reduction in energy utility but also may bring reduction in the global ΔTmin. Thus in the current state-of-the-art pinch technology applications the capital/energy trade-off in energy system synthesis should be readjusted after each process change. This process of readjusting after each change is a real challenge due to the combinatorial nature of the whole process-utility system. In addition the change in the driving force distribution due to stream-specific ΔTmin and all possible combinations of process changes would have an affect on the utility level selection and quantities. Simply the problem of finding optimal utility system and heat recovery system driving force distribution is interlinked and multi-dimensional in which process changes and stream-specific ΔTmin selection are competing for optimal selection of utilities, optimal process conditions and optimal energy recovery system synthesis as well as optimal utility system synthesis. Pinch technology most widely used in process industry falls short in solving, systematically and without iteration, the first part of the problem which is the finding of the energy targets ahead of the process and utility systems synthesis due to all possible combinations of process conditions changes and different practical stream-specific ΔTmin.

In this paper we introduce a systematic method and software for energy targeting under varying driving force distribution conditions in heat recovery systems.

A Holistic Approach to Capturing and Sustaining Energy Savings

Jose Desouza and Piyush Shah. Aspen Technology, 2500 CityWest Blvd, Houston, TX 77494

Process manufacturers are most vulnerable to the recent rise in energy costs. According to the U.S. Department of Energy, the percentage of operating costs attributable to energy range from about 7% to over 30% for the process industries . The result is eroding profit margins and increased pressure from shareholders to take action. It therefore comes as no surprise that energy costs and energy reduction initiatives are among the top business issues cited by company executives.

This paper will illustrate the value of taking a holistic approach to energy reduction versus deploying single point solutions. A holistic approach enables manufacturers to capture the most opportunities to increase bottom line profitability, and ensure long-term sustainability. Such a strategy addresses four key focus areas:

• Invest wisely • Use less • Buy cheaper • Generate and distribute more efficiently

In a holistic approach, an energy efficiency improvement program is targeted at a site. Improvement projects with documented benefits are considered alongside other site operations and capital investment opportunities. Approved projects are then sequenced and coordinated to ensure rapid improvement, knowledge and technology transfer and sustainability. The types of projects implemented will impact the design and operation of process and utilities systems and range widely in scope and capital investment.

This approach to reducing energy costs for process industry manufacturers will be demonstrated by means of a case-study on a typical petrochemicals manufacturing complex. This case-study compares the results of a holistic energy efficiency improvement program to the implementation of several point solutions in an uncoordinated fashion. We will show how the process design objectives of capacity increase and optimum energy efficiency can both be achieved while avoiding unnecessary capital investment. We will also show that if the implementations of operational support tools are not carefully coordinated, they can end up pulling in different directions, resulting in sub-optimal overall performance.

Energy Storage by Modeling a Cooling System

Fatemeh Abniki, National Petrochemical Company,Ghadir Group, Tehran, Iran and Ehsan Bakhshi, R&D Center of National Petrochemical Company, Tehran, Iran.

The sensible heat storage, larger amounts of energy per unit volume can be stored by utilizing the latent heat of a phase change . In this case, during the daytime hours the incident sunlight shining onto the building is stored as the salt melts. Then at nighttime ,as the salt solidified, the released energy can heat the building .Recently, latent heat " cold storage ' has been proposed as a means of leveling air conditioning loads in hot climates. One scheme, which is being commercialized, uses latent heat to store waste heat from an automobile engine so that it may be used to preheat the engine in cold climates following being off for more than several hours. It is evident that in a typical packed bed regenerator, far greater amounts of energy can be stored if the packing were replaced with material which changes phase and stores energy as latent heat. In this case , the heat storage medium is some means of containing or supporting the phase change materials, PCM, must be provided since the material can flow when it is in the liquid state .Models for the phase change regenerator, PCR ,are complex , involving the solution of the classical Stefan problem for phase change heat storage on the scale of the PCM, coupled with the heat balance equation for the scale of the PCR packed bed. The model presented accounts for convection, heat accumulation in the fluid ,heat losses through the wall of the PCR and energy storage by PCM. A PCM consisting of water in spherical support polypropylene (P.P) is used in a laboratory scale PCR to verify the model. Experiments with heated or cooled air passing through the PCR are described. The measured outlet temperature compares qualitatively with the model predictions. With the importance of heat recovery echnology to reduce energy consumption, there is to develop an understanding for the PCR operation. We have presented a computational scheme for the prediction of temperature profiles within the bed and the outlet fluid temperature . Parametric studies have verified that for all practical purpose , the ideal PCR is approached for: $(3Bi)/St<5*10$; $Pe > 40$; $St > 50$ The computational model which was developed is comprehensive and robust. It is capable of assessing the effect of operating and design parameters and performance of commercial-size PCRs. Comparisons made between the outlet fluid temperature which was measured experimentally to that predicted by the model indicate that, in the small insulated bench-scale apparatus, there is a signification amount of heat loss. As much as50% of the energy stored in the experimental unit can be lost to the surrounding via the end plates.

Web Page: www.nipc.net

Thermodynamic and Kinetic Study on Dimethyl Ether Steam Reforming

Cong Li and xinghu Li. Dept. of Automobile Engineering, Beijing University of Aeronautics and Astronautics,china, beijing,china, 0086100083, China

Based on a thermodynamic analysis, the effects of temperature, the ratio of steam to dimethyl ether (DME) on thermodynamic equilibrium compositions in the DME steam reforming system were obtained. The ratio of steam to DME to be 3 or more and low temperature is favorable for the reaction system equilibrium compositions.A system of DME steam reformation reaction for experiment and measurement was designed. Using this system, kinetics of DME steam reforming was investigated. The effects of temperature, the ratio of steam to DME and W/F which is the ratio of quality of catalysts and DME velocity of flow on conversion of DME were analyzed. Kinetics expressions parameters of Arrhenius style simplified model were determined on the data obtained from experiments.

Hydrogen Production Via Gasification of Solid Carbon Fuels

Marco J. Castaldi and Heidi Butterman. Earth & Environmental Engineering (HKSM), Columbia Univeristy, 500 West 120th street, 926 Mudd Building, New York, NY 10027

This presentation will compare experimental and simulation results of an integrated catalytic combustion, solid carbon gasifier to produce hydrogen and electricity. The technology, which is a form of steam reforming, focuses on the combining of catalytic combustion with gasification to generate H2 and CO from a carbon source such as biomass while yielding a CO_2 sequestration ready stream. The H_2 can either be separated from the CO or remain as a mixed stream for use as an IGCC turbine fuel. An Aspen® Plus simulation shows that a steam to carbon ratio of 1.5 provides a hydrogen output of 1.15 kg/hr while generating about 14 kW of electricity (for every one kmol/hr of carbon fed to the reformer) from a SOFC using the portion of the CO generated that was not needed to drive the reforming reactions. Additionally, recycling up to 25% of CO_2 into the reformer produces about 15% more hydrogen, while using 20% less CO for combustion to drive the gasification reactions. Because of the Boudouard reaction an extra 32% electrical capacity (4.5 kW per kmol/hr of carbon) can be generated from an SOFC operating on the CO not used for combustion. The advantage of coupling catalytic combustion with the gasification process is that it enables the heat producing combustion reactions to operate at lower temperatures thus decreasing NO_x and resulting in more energy efficient conversion. A portion of the carbon monoxide is catalytically combusted to carbon dioxide with either stoichiometric amounts of oxygen or air. If combusted with oxygen a pure carbon dioxide stream is produced that can be sent for sequestration with minimal oxygen usage. A series of experiments were done to compare against the ASPEN® simulation results. The experiments were mainly focused on biomass gasification to convert carbon to useful forms of energy while addressing environmental concerns. Using Thermogravimetric Analysis (TGA) we studied the biomass decomposition rates and formation of char residue. Several biomass sources were analyzed individually to enable comparisons between each species and help adjust the ASPEN® parameters to better reflect biomass conversion in general. The species tested were various pulverized woods and grasses that were comprised of maple, poplar, pine, spruce, oak, green Douglas fir, pine needles, maple bark, alfalfa, switchgrass and cordgrass. Each specimen feedstock was slowly heated at a rate of 10°C min^{-1} in a nitrogen-steam atmosphere. The woods and grasses had similar TGA decay curves, resulting in three main decomposition regions. Not until the third decay region, in which the temperatures reached between 900°C -1000°C, was the biomass decomposition complete. In addition, Bituminous coal particles (38, 45, 53, 63 and 106 μm) were tested under the same conditions to ascertain major differences between varieties of carbon feedstocks. Compared to the biomass decomposition thermograms, most of the coal samples were still experiencing mass loss at 1000°C and thus did not exhibit the plateau observed for biomass. To determine the feasibility of enhanced hydrogen production from gas recycle, differing amounts of carbon dioxide to nitrogen gas ratios were introduced (0%, 25%, 50%, and 75% CO_2) and coupled with online gas chromatography. Four major products were quantified, H_2, CO, CH_4 and N_2, to determine a carbon dioxide recycle concentration that could optimize the rate of gasification of the solids and simultaneously increase the yield of H_2 and CO. Verification of the numerical simulations was experimentally observed under certain concentrations for solid carbon (i.e. C) gasification at atmospheric pressure. Finally, the residuals of the biomass test samples were analyzed using Atomic Adsorption Spectrometry. A significant quantity of potassium was identified in the grass chars which is likely responsible for observed corrosive behavior of the grasses during gasification in the TGA. The high alkaline content of many grasses is an important consideration in the choice of an appropriate carbon-neutral biofuel and likely will require pre-processing necessary to minimize corrosion of the process equipment. The biomass feedstocks vary in their concentrations of active constituents such as potassium, calcium, magnesium, phosphorus and sulfur that can pose a major problem to downstream equipment.

Keywords: biomass, waste-to-energy, gasification, catalytic combustion, hydrogen generation

AUTOTHERMAL PARTIAL OXIDATION OF METHYL ACETATE IN A CATALYTIC BED REACTOR

Betty Nguyen and Corey Leclerc**
Department of Chemical Engineering
McGill University
3610 University St
Montreal, QC H3A 2B2 CANADA
corey.leclerc@mcgill.ca

** To whom correspondence should be sent

Key Words: biodiesel, hydrogen, rhodium, methyl acetate

Prepared for Presentation at the 2007 Spring Meeting, Houston, TX, April 22-26, 2007

AUTOTHERMAL PARTIAL OXIDATION OF METHYL ACETATE IN A CATALYTIC BED REACTOR

Betty Nguyen and Corey Leclerc

Department of Chemical Engineering, McGill University

Montreal, QC H3A 2B2

ABSTRACT

Synthesis gas constitutes the main source of hydrogen for fuel cells. The lower temperatures at which existing fuel cells operate in contrast to those in internal combustion engines considerably reduce the emission of nitrous oxides. Previous research has shown that compounds as diverse as methane and gasoline can be used to produce synthesis gas. The use of a renewable fuel such as biodiesel for such a process is desired, since its derivation from biomass results in a closed carbon cycle and the net carbon dioxide emission from its use is significantly lower than that from the use of fossil fuels. By combining renewable fuels with fuel cells, it is possible to reduce the emissions of carbon dioxide and nitrous oxides.

Methyl acetate has been chosen as a model compound for biodiesel, in order to isolate the methyl ester functional group from the hydrocarbon chain. Methyl acetate is fed into a millisecond reactor along with air, where it undergoes catalytic partial oxidation (CPO) to produce a mixture of hydrogen, carbon monoxide, carbon dioxide, water, and some hydrocarbons. Several catalysts were deposited onto washcoated, α-alumina foam monoliths. The conversion of the methyl acetate and the selectivity of various product species are studied for different feed flow rates and carbon to oxygen feed ratios.

INTRODUCTION

Current energy research has focused on developing alternatives to combustion processes as well as alternatives to fossil fuels. Combustion produces nitrous oxides that are major contributors to photochemical smog. Fossil fuels are projected to run-out within a finite time frame. Additionally, their use leads to the accumulation of carbon dioxide in the troposphere, which contributes to the greenhouse effect and global warming.

Fuel cells are an alternative to combustion. Their operating temperature is low compared to combustion processes leading to the absence of nitrous oxides in the exhaust. Additionally, fuel cells obtain higher efficiencies, since they do not rely on the transfer of heat and are thus not limited by the Carnot efficiency. Fuel cells should obtain a higher "well-to-wheel" efficiency than combustion engines.

Biofuels, such as biodiesel, have many potential benefits over fossil fuels. Since they require carbon dioxide for growth, ideally carbon dioxide produced from energy production will be used to grow more biomass. This will lead to a closed carbon cycle where carbon dioxide does not accumulate in the troposphere preventing global warming. Additionally, biofuels hold many potential socio-political benefits such as domestic energy production ending the reliance on foreign governments and a revigoration of the rural farmer who would become a primary energy producer for the country.

Schmidt et al. [ref] have demonstrated a reactor that can convert biodiesel to hydrogen via catalytic partial oxidation (CPO). Their process converts >90% of the biodiesel to hydrogen with a selecitivity of ~80%. The process takes place in milliseconds leading to a reactor that is on the order of milliliters to produce enough hydrogen for an automobile. A combination of rhodium and cerium was found to be the best catalyst for this process.

Biodiesel is comprised of a methyl ester group and a long chain alkane or alkene. Previous studies have demonstrated the ability of the CPO process to convert long chain alkanes and biodiesel into hydrogen. In this work, we split biodiesel into two functional groups and only consider the methyl ester functional group. We perform the CPO of methyl acetate as a first study of the methyl ester functional group. Methyl acetate is the simplest methyl ester where the C_{16}-C_{20} carbon chain that is found in biodiesel is substituted with a methyl group. The goal is to gain some insight into the conversion of the ester group and determine the ability of our catalysts to convert the ester. We have used several catalysts including rhodium and rhodium-cerium which have shown high activity for biodiesel, platinum which has shown high activity for methane, and palladium and copper which have shown high activity for methanol.

EXPERIMENTAL

Apparatus

A schematic of the reactor setup is shown in Figure 1. The methyl acetate is stored in a tank pressurized to 3 psig with nitrogen. It is introduced to the reactor through an automotive fuel injector. As the methyl acetate enters the reactor, it is mixed with air and vaporized on the reactor walls or on blank monoliths upstream from the catalyst. In this region, the walls are heated with a resistive heating tape. The air is fed into the reactor through a sidearm, and its flowrate is controlled by a Brooks Smart Series 5850S mass flow controller. The outer reactor wall temperature is recorded with a K-type thermocouple inserted between the heating tape and the outer reactor wall.

The methyl acetate and air mixture reacts as it passes through the catalyst setup further downstream in the reactor. The catalyst-supporting monolith is positioned between two blank monoliths, which serve as axial heat shields. The three monoliths are wrapped in an alumino-silicate paper (Fiberfrax) to prevent reactant bypass. Furthermore, a K-type thermocouple is inserted through the center of the downstream heat shield, to measure and record the catalyst's back face temperature. The product gas stream is vented through a reactor sidearm downstream of the catalyst setup. The fuel injector and the mass flow controller are both controlled and the thermocouples' data are recorded with LabVIEW 7.1 (National Instruments).

The reactor ignition is achieved by first heating the catalyst with a Bunsen burner flame to about 200°C. The heating tape around the upper section of the reaction is switched on, and its power output is controlled by a variable transformer. The heating tape output setting is set for a target outer reactor wall temperature of 100°C ± 40°C, to ensure that the temperature in that section of the reactor would be above the methyl acetate's normal boiling point of 57°C. The methyl acetate and air are then fed and when light-off occurs a large increase in back face temperature is observed. The Bunsen burner flame is then removed and the reactor is wrapped in an Isofrax blanket to prevent radial heat losses. The reactor is then left to reach steady state operation and the product gases are sampled with a gas tight syringe. The gas sample is injected into a dual column Agilent 6890N gas chromatograph (GC). The GC is equipped with a HP-PLOT Q column to separate the carbon dioxide and various oxygenates present in the gas sample, and a HP-PLOT molesieve column to separate the permanent gases (nitrogen, oxygen, carbon monoxide, methane, hydrogen). The hydrogen content of the reactor product stream is determined by measuring the area of the gas chromatograph's corresponding inverted signal peak. In contrast, the water content in the product stream is calculated by closing the reactor's elemental oxygen balance. The elemental carbon balances are calculated and close to within ±15%. The balances were not completely closed due to the formation of solid carbon either in the catalyst or downstream in the reactor. The formation of carbon in the reactor is discussed later on.

Figure 1. Reactor Configuration

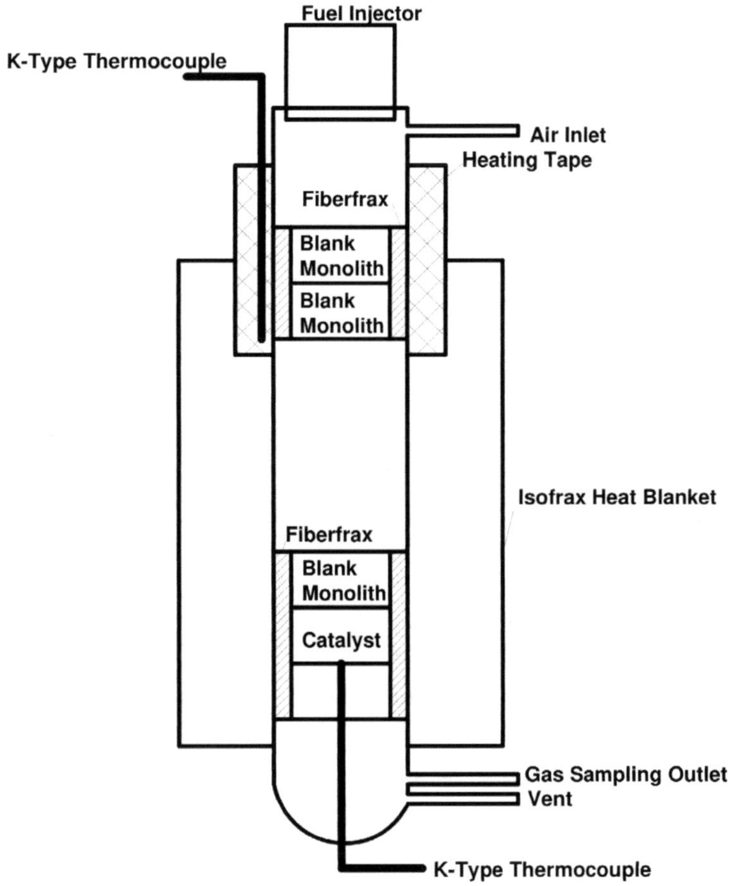

Catalyst Preparation

Cylindrical α-alumina foam monoliths (Vesuvius Hi-Tech) with a diameter of 17 mm, pore density of 80 pores per inch, and porosity of 0.83 served as the support material for the catalytic metal. The monoliths are first washcoated with γ-alumina (Degussa) for a target washcoat loading of 5 wt% in order to increase the surface area and decrease the pore diameter. The monoliths are then allowed to dry overnight. An aqueous solution of the desired metal precursor (rhodium (III) nitrate, cerium (III) nitrate hexahydrate, copper nitrate, palladium nitrate, and hexachloroplatinic acid) is prepared and pipeted onto the monolith. The rhodium-ceria catalyst is a mixture of 2.5 wt% Rh and 2.5 wt% Ce, whereas the other catalyst loadings are ~5 wt%. The catalyst is dried overnight and calcined in air at 600°C for 8 hours.

Reactor Performance

The catalyst's suitability for the CPO of methyl acetate was evaluated on the basis of the resulting reactant conversions and the product selectivities. The conversion of a reactant *i* is defined as the ratio of the amount of reactant *i* which was consumed in the reactor, over that which was fed to the reactor:

$$X_i = \frac{F_{i,in} - F_{i,out}}{F_{i,in}}$$

where $F_{i,in}$ is the molar flow rate of reactant *i* entering the reactor and $F_{i,out}$ is the molar flow rate of reactant *i* exiting the reactor.

The selectivity of element *i* to form product *j* ($S_{i,j}$) is calculated by:

$$S_{i,j} = \frac{\nu_{i,j} F_{j,out}}{\sum_k \nu_{i,k} F_{k,out}}$$

where $\nu_{i,j}$ is the stoichiometric number of atoms of element *i* in product *j*. The denominator is the summation over all species, *k*. In this work, the selectivities for hydrogen, water, carbon monoxide, and carbon dioxide are shown, since they are the major reactor products. Since the hydrogen and water selectivities are both calculated on the basis of atomic hydrogen and the carbon dioxide and carbon monoxide selectivities are based on atomic carbon, the *i* subscript in the selectivity variable is omitted. The selectivities pertaining to a single element (carbon or hydrogen) presented in this work may not add up to 100%, since not all of the products are presented.

The effects of the C/O feed ratio on the reactor temperature and the product composition are studied in this work. These were selected as the independent variables since previous research on the partial oxidation of alkanes has shown that the catalytic metal loading and the catalyst's support material have little effect on the conversions and selectivities, regardless of the fuel or catalyst [2].

The gas hourly space velocity is given as:

$$GHSV = \frac{Q_{total,STP}}{\varepsilon V_{monolith}}$$

where $Q_{total,STP}$ is the total volumetric flow rate entering the reactor at standard temperature and pressure, ε is the monolith void fraction, and $V_{monolith}$ is the volume of the monolith.

The C/O feed ratio is defined as the number of carbon atoms entering the reactor divided by the number of oxygen atoms entering the reactor. Therefore pure methyl acetate has a C/O ratio of 1.5 while syngas has a C/O ratio of 1.0.

It is important to note that the overall partial oxidation of methyl acetate (1) is in fact an endothermic reaction, as it was reported to be the case for ethanol [3-4].

$C_3H_6O_2 + ½ O_2 \rightarrow 3 CO + 3 H_2$ $\Delta H = +78.4$ kJ/mol (1)

This requires the input of a small amount of excess oxygen in order to have some combustion (2) occurring in the reactor, thus supplying the heat for sustaining the autothermal operation of the reactor.

$C_3H_6O_2 + 3 ½ O_2 \rightarrow 3 CO_2 + 3 H_2O$ $\Delta H = -1496$ kJ/mol (2)

RESULTS

Feed Stoichiometry

Figure 2 illustrates the effect of a varying C/O feed ratio on the back face temperature (a), methyl acetate conversion (b), hydrogen selectivity (c), water selectivity (d), carbon monoxide selectivity (e) and carbon dioxide selectivity (f). For these graphs the gas hourly space velocity was kept constant at 143,300 h^{-1}.

The back face temperatures for both catalysts decrease continuously in response to increases of the feed ratio. However, the RhCe catalyst shows a higher back face temperature than the Rh catalyst over the entire range of feed ratios studied. It can also be observed that the back face temperature difference decreases as the feed ratio increases. For a C/O feed ratio of 0.75, there is a difference greater than 200°C between the catalysts' back face temperatures. At a C/O feed ratio of 0.87, the back face temperatures range from 776°C to 790°C.

Although the Rh catalyst demonstrates the highest methyl acetate conversion at a C/O ratio of 0.75, both catalysts give out very similar methyl acetate conversions for most of the range of feed ratios studied. The methyl acetate conversion for both catalysts decreased continuously with increases in C/O feed ratios. The Rh catalyst's methyl acetate conversion dropped from 92% to 56% and the RhCe catalyst's methyl acetate conversion declined from 79% to 53% as the C/O feed ratio increased from 0.75 to 0.87.

A decrease in hydrogen selectivity is observed for both catalysts in response to an increase in feed ratio. The Rh catalyst's hydrogen selectivity dropped from 32% to 12% over the range of C/O feed ratios whereas the RhCe catalyst experienced a reduction in hydrogen selectivity from 36% to 16%. Overall, the RhCe catalyst seems to produce higher hydrogen selectivities than the Rh catalyst. In contrast to the hydrogen selectivity, the water selectivity increases as the feed ratio increases. The Rh catalyst's water selectivity rose from 27% to 46% whereas the RhCe catalyst's water selectivity increased from 40% to 55%.

Correspondingly, the carbon monoxide selectivity shows a decrease as a result of an increase in feed ratio. The carbon monoxide selectivity observed with the Rh catalyst ranges from 36% to 50% while for the RhCe catalyst, it ranges between 59% and 77%. Similarly to the hydrogen selectivity, higher carbon monoxide selectivities appear to be experienced with the RhCe catalyst. Conversely, the carbon dioxide selectivity remained constant over the range of feed ratios for both catalysts. The RhCe catalyst shows a carbon dioxide selectivity of around 16% while the Rh catalyst experiences a carbon dioxide selectivity of about 13%.

Figure 2: Back Face Temperature, Reactant Conversion, and Product Selecitivity for Varying C/O Ratios

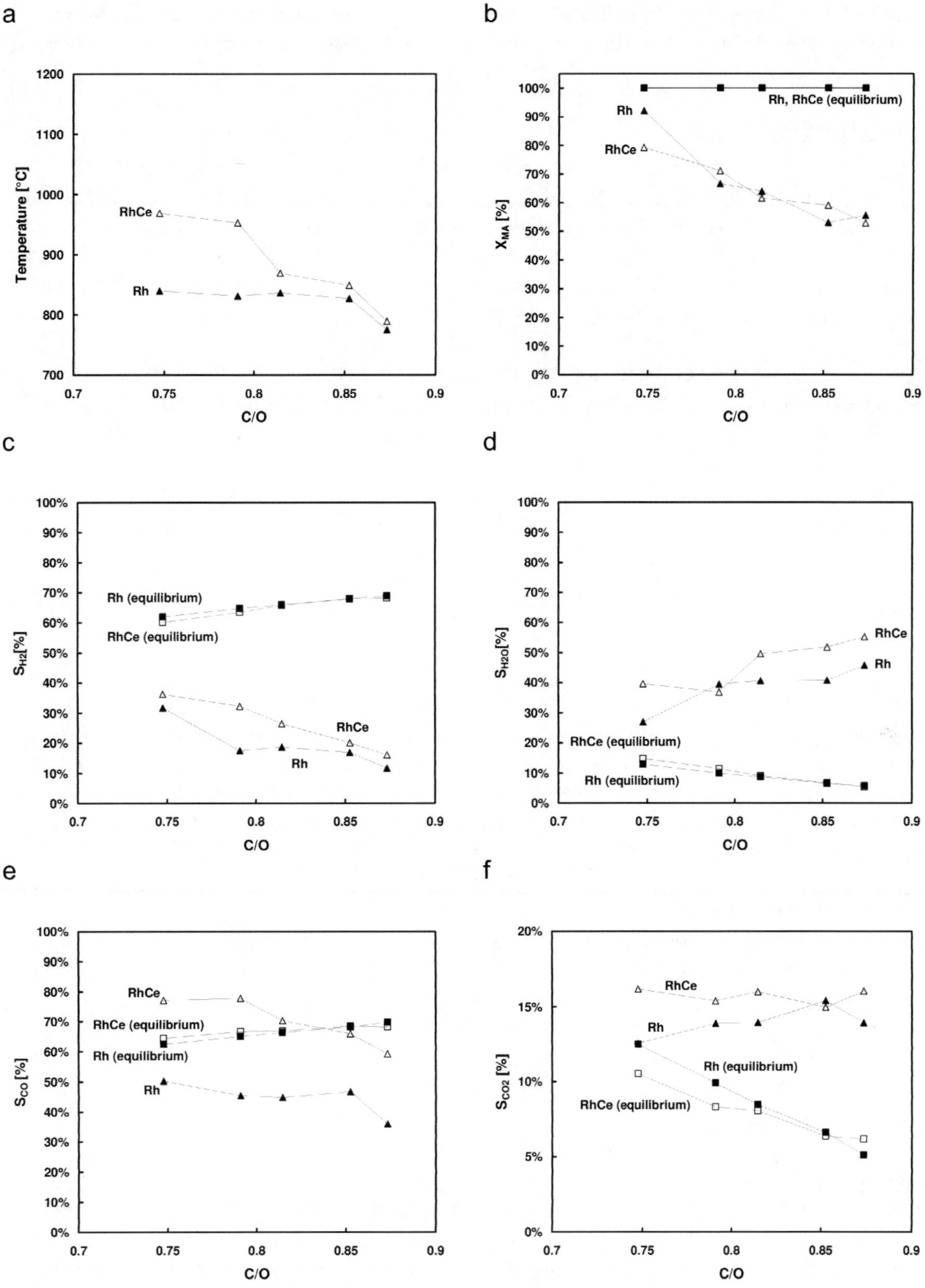

The other byproducts observed in the GC signal were ethane, ethylene, acetylene, methanol, acetaldehyde, and methane. Their corresponding selectivities do not exceed 7% and were relatively constant throughout the range of C/O feed ratios examined, regardless of the catalyst used. For most byproducts, the RhCe catalyst exhibits the greater selectivity. The Rh catalyst showed carbon formation whereas the RhCe catalyst did not. However, catalyst degradation was never observed.

Comparison to Biodiesel Results

A series of experiments at a total flow rate of 4 slpm (GHSV = 127,300 h^{-1}) were performed in order to compare the resulting fuel conversion and product selectivities to those pertaining to the partial oxidation of biodiesel from previous literature [ref], with a RhCe catalyst. Figure 4 shows the evolution of the back face temperature (a), the methyl acetate conversion (b), the hydrogen selectivities (c), and the carbon selectivities (d).

Similarly to the experiments at constant gas hourly space velocity, the back face temperature continuously decreased from 896°C to 821°C as the C/O feed ratio increased from 0.75 to 0.87. Also, the methyl acetate conversion dropped from 81% to 62% over the same range of feed ratios. While the water selectivity increased from 36% to 43% with the increase in feed ratios, the hydrogen selectivity decreased from 39% to 18%. The carbon dioxide selectivity remained mostly constant around 14-15% whereas the carbon monoxide selectivity decreased from 78% to 64%.

DISCUSSION

Catalysts

The rhodium and rhodium-cerium catalysts were the only catalysts to light off and sustain autothermal operation. The platinum and copper catalysts did not light off. The color of the copper catalyst changed indicating a change in oxidation state, but it could not sustain light off. The palladium catalyst did light off, but coke formation led to deactivation in less than 20 minutes. Since the rhodium and rhodium-cerium catalysts were the only catalysts to light off, only their results were presented.

Rhodium versus Rhodium-Cerium

The results show little deviation between the rhodium and rhodium-cerium catalysts. Previous research with oxygenated hydrocarbon CPO has shown that the addition of cerium increases the conversion of fuel and selectivity to hydrogen of a rhodium catalyst. In this work, the difference between the two catalysts is quite small. This indicates that the surface reaction may not be as important for the methyl acetate as it was for ethanol. It is expected that investigation of the gas phase reactions will show that methyl acetate is more reactive homogeneously which diminishes the importance of the catalyst.

Comparison to Biodiesel

The previous research of biodiesel CPO showed that a conversion of ~95% with a selectivity to hydrogen of 80% is possible. However, for methyl acetate, the conversion and selectivity are much lower than biodiesel. The hydrogen selectivity reached 40% for methyl acetate as a maximum.

CONCLUSIONS

Rhodium and rhodium-cerium are the best catalysts of those investigated for the CPO of methyl acetate, the simplest methyl ester. The conversion of methyl acetate and selectivity to hydrogen are far below typical values for the CPO process. The reactions occurring on the catalyst will be important in choosing a more suitable catalyst for the conversion of methyl acetate. In order to do so, a better understanding of the reactions occurring in the gas phase will be important to give an idea of what molecules are adsorbing and subsequently reacting on the catalyst surface.

ACKNOWLEDGEMENTS

The authors thank the Natural Sciences and Engineering Research Council of Canada for research funding. BNT thanks the Eugenie Ulmer Lamothe Fund for funding her graduate studies.

REFERENCES

1) Subramanian R and LD Schmidt. 2005. Renewable fuels from biodiesel by autothermal reforming. Ang. Chem. Int. Ed. 44, 302-305.
2) Bodke AS, SS Bharadwaj, and LD Schmidt. 1998. The effect of ceramic supports on partial oxidation of hydrocarbons over noble metal coated monoliths. J. Catal. 179, 138-149.
3) Salge JR, GA Deluga, and LD Schmidt. 2005. Catalytic partial oxidation of ethanol over noble metal catalysts. J. Catal. 235, 69-78.
4) Deluga GA, JR Salge, LD Schmidt, and XE Verykios. 2004. Renewable hydrogen from ethanol by autothermal reforming. Science. 303, 993-997.

Thermodynamic Analysis of a Microbial Cell: a Process Synthesis Approach

G. A. Darko[1], D Hildebrandt[2], and D Glasser[2]. (1) Chemical and Metallurgical Engineering, Wits University, Johannesburg, South Africa, (2) Chemical and Metallurgical Engineering, Wits University, Johannesburg, South Africa

Abstract

In this contribution, it is shown that with the right kind of data measurement, it is possible to extract vital information (from an experimental measurement) that could be used for the analysis and design of a process. Using the three process synthesis tools (mass, energy and entropy balances), thermodynamic properties such as enthalpy of formation (Hfcell) and heat production (qcell) of a bacterium cell are estimated from a chemostat experimental measurement reported by Collet et al (2004). Gibbs free energy of formation (Gfcell) and maintenance free energy (mE) of the cell are also estimated. The values of Hfcell and mE agree well with the vast amount of literature present (von Stockar et al, 2005; von Stockar and Liu,1999; Tijhuis et al, 1993; Russel,1986; Roels,1983). The values for Gfcell and qcell were not far off literature values either (von Stockar et al, 2005; Russel, 1986). The results have allowed us to be able to predict possible thermodynamically feasible reaction for a given substrate and product (s); hence setting process targets (in this case, predicting maximum hydrogen production).

Key words: thermodynamics, fermentation reactions, process targets

References

Collet, C., Adler, N., Schwitzguébel, J. P. and Péringer, P. 2004. Hydrogen Production by Clostridium thermolacticum during continuous fermentation of lactose. International Journal of Hydrogen Energy, 29, 1479 – 1485

Russell, J. B., 1986. Heat production by ruminal bacteria in continuous culture and its relationship to maintenance energy. Journal of Bacteriology, 168, 2, 694 – 701

Roels, J. A., 1983. Energetics and Kinetics in biotechnology. Elsevier biomedical press, Amsterdam.

Tijhuis, L., van Loosdrecht, M.C.M., Heijnen, J.J., 1993. A thermodynamically based correlation for maintenance Gibbs energy requirements in aerobic and anaerobic chomotrophic growth. Biotechnol. Bioen., 42, 509-519

von Stockar, U., Liu, J-S., 1999. Does microbial life always feed on negative entropy? Thermodynamic analysis of microbial growth. Biochem. Biophys. Acta. 1412, 191-211.

von Stockar, U., Maskow, T., Liu, J., Marison, I. W., Patiño, R., 2005. Thermodynamics of a microbial growth and metabolism: An analysis of the current situation. J. Biotechnol.,DTD 5.

TECHNICAL AND ECONOMIC ANALYSIS OF HYDROGEN PRODUCTION USING OCEAN THERMAL ENERGY CONVERSION

C. B. Panchal **
Argonne National Laboratory
Argonne, IL 60439
cpanchal@anl.gov

Peter Pandolfini
The John Hopkins University/Applied Physics Laboratory, Laurel, MD
Peter.Pandolfini@jhuapl.edu

William Kumm
Arctic Energies, 511 Heavitree Lane, Severn Park, MD 21146
whkwhoswhoworld@aol.com

** To whom correspondence should be sent

Key Words: hydrogen, ammonia, ocean thermal energy conversion, renewable energy

Prepared for Presentation at the 2007 National Meeting, Houston, TX April 22-26, 2007

Unpublished

TECHNICAL AND ECONOMIC ANALYSIS OF HYDROGEN PRODUCTION USING OCEAN THERMAL ENERGY CONVERSION

C.B. Panchal, Peter Pandolfini, and William Kumm

Abstract

The world's oceans are the largest collector and storage of solar energy and have an enormous potential to supply the growing worldwide energy demand. This analysis examines the benefits of at-sea production of ammonia, as the hydrogen carrier, with co-production of desalinated fresh water. Ammonia offers significant advantages for hydrogen transport and storage, as compared to other options being evaluated such as hydrides, compressed or liquefied hydrogen. The potential use of ammonia as the hydrogen delivery systems is further supported by the established surface and ocean transportation systems for delivering these two commodities to the coastal states of the United States, with a further, enormous potential for the world market. Achieving this goal of ammonia-based distributed power generation may take some time; however, there is a significant near-term market opportunity for hydrogen derived from renewable non-carbon sources in the petrochemical and refining industries in California and the Gulf of Mexico coastal states and for production of non-carbon ammonia as fertilizer.

Introduction

The National Hydrogen Roadmap highlights the development of advanced renewable energy methods that do not emit carbon dioxide. In 2005, the US Department of Energy developed a hydrogen cost goal of \$2 to \$3/kg (in 2005 dollars) of hydrogen delivered by 2015. In comparison, the steam reforming-based hydrogen cost is about \$1.1/kg at the natural gas cost of \$5/MBtu. Coal gasification combined with carbon sequestration may cost over \$2/kg hydrogen. Distributed hydrogen production using electricity is dependent on local electric rates (Ivy, 2004) and the cost of hydrogen can vary from \$2-\$3/kg at electric rates of \$50/MWh to greater than \$5/kg for \$100/MWh. Potential options for transportation and storage of hydrogen are hydrides, compressed or liquefied hydrogen. These options will require the development of new infrastructure for reaching hydrogen to end users. Avery (1988) showed that ammonia, as the hydrogen carrier, offers an alternate carrier option, with high energy density as compared to these other options. The technology and infrastructure for transportation (sea and land) and storage of ammonia are well established. Therefore, ammonia is a good candidate as the hydrogen carrier for the near-term hydrogen economy. Ammonia can be readily cracked to produce hydrogen on site (Kacker and Kordesch, 2003). A report by Risø National Laboratory (Duijm, Markert, and Paulsen; 2005) provides a comprehensive analysis of ammonia as the transportation fuel. A DOE study (Thomas and Parks, 2006) concluded that "*Ammonia may be considered as a potential hydrogen carrier for hydrogen delivery and off-board storage, such as refueling stations and for stationary power applications.*"

Ammonia is the backbone of fertilizer production with world's production of about 140 million metric ton per annum (MTA). Ammonia is currently produced from natural gas (NG), coal, refinery residuum, and naphtha. NG is the primary feedstock for ammonia manufacturing in the USA while naphtha and NG are used in Europe. Other countries use coal and refinery heavy residuum with significantly higher emission of carbon dioxide as compared to ammonia produced from NG. The cost of ammonia has risen from about $150 per metric ton (MT) when NG prices were low to greater than $350/MT (FOB Gulf Coast) in 2006 (Kubic, 2006). NG consumption in manufacturing of ammonia is about 36.3 MBtu/MT. At a NG cost of $6/MBtu, the "*raw material*" cost of NG alone would be $218/MT, which explains why many of the US domestic manufacturing plants have been mothballed. Coal-based ammonia cost in the USA would be $237/MT at $30/ton of coal or coke and electric rate of $50/MWh (Markey 2006). An estimated cost of ammonia using nuclear energy to produce hydrogen from electrolysis is $181/MT and $263/MT for a high-temperature gas-cooled reactor (HTGR) and a pressurized-water reactor (PWR), respectively (Kubic 2006). These cost estimates are based on nuclear heat available at $5.5/MWh.

Among renewable energy options, ocean thermal energy provides a base-load supply of solar energy. The general perception is that ocean energy is limited to only tropical countries by providing electric power via cable. Therefore, the full potential of ocean energy for at-sea production of ammonia as the hydrogen carrier and desalinated water production has not been adequately evaluated. With the focused R&D by the US Department of Energy (DOE) and the Maritime Administration (MarAd) of Department of Commerce in the 1970's and 1980's, the technical risks and design uncertainties have been reduced to an acceptable level for commercial deployment of Ocean Thermal Energy Conversion (OTEC) plantships. However, due to cheaper fossil energy in 1970's and 1980's, further development of OTEC diminished to low-level developments. The present energy, environmental and economic conditions provide a *Window of Opportunity* for at-sea OTEC plantships for the co-production of ammonia and fresh water, and therefore, the near-term deployment of multiple OTEC plantships and their associated transport tanker fleets.

The purpose of this design analysis is to evaluate the technical and economic viability of the OTEC plantships for co-production of ammonia and fresh water. This analysis is based on a detailed OTEC ammonia pilot plantship design performed by The Johns Hopkins University Applied Physics Laboratory (JHU/APL) in collaboration with industrial partners (George and Richards, 1980). In this design study, an improved power system based on compact brazed aluminum heat exchangers and water production plant is incorporated into the JHU/APL plantship. The original work-breakdown structure (WBS) is updated to reflect the updated design and fresh water production. The cost estimates are updated using the *Chemical Engineering* cost index and recent quotes of major subsystems. The cost of ammonia is derived from the ***H2A*** Excel-based economic model.

OTEC Plants – Technology Status

The OTEC system produces electric power using the temperature difference of warm surface and deep-ocean cold waters. In 1970's and 1980's, DOE, jointly with the industry, pursued large floating OTEC plants for generating electric power and transporting it to the land via submerged cables. Figures 1 shows a conceptual design of the TRW 100 MWe OTEC plant. In a later stage of the OTEC program, the focus shifted to smaller plants (10 to 40 MWe) which can be located on the land or in shallow waters as shown in Figure 2 of the GE's tower design concept. These small shore-based plants were intended for the Island market (Steven, Genes, and Panchal; 1984). The economics of small OTEC plants were shown to be promising for hybrid OTEC plants for co-production of power and fresh water for regions where RO-based desalination is being used (Panchal and Bell, 1987; Rabas, Panchal and Genens, 1990). An economic analysis showed that OTEC plantships are

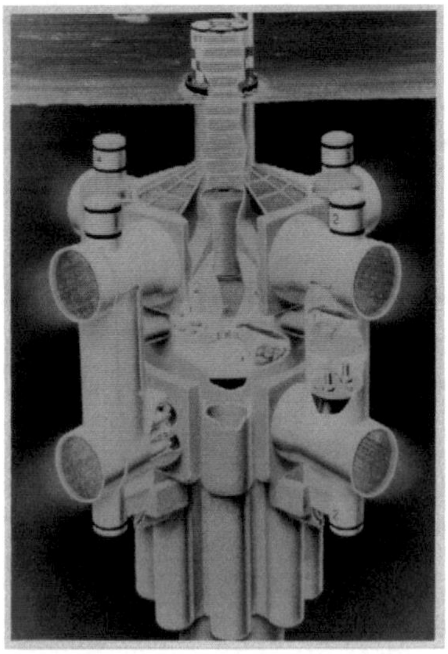

Figure 1. TRW 100 MWe OTEC semi-submerged floating plant concept

competitive with the petroleum-based power generation, when crude oil prices are greater than \$35/barrel and desalinated water can be marketed for \$0.8/m^3 (\$3/1000 gallon). Although, the design analyses show promising economics, the industry's interests in OTEC did not grow when the energy costs were relatively low. The present energy costs, environmental concerns and energy-supply security have prompted renewed interests in the OTEC technology.

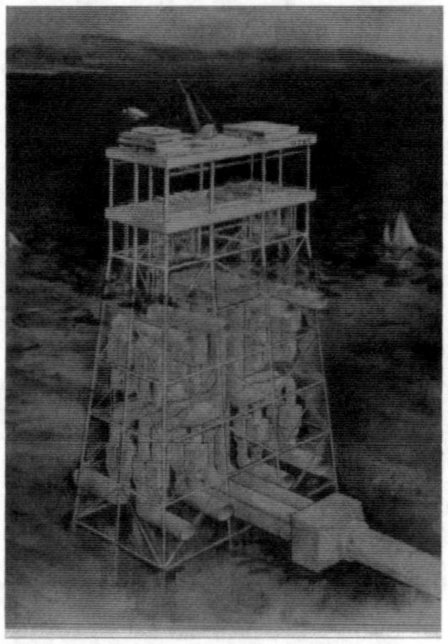

Figure 6. General Electric
nt

OTEC Plantships

In a visionary approach Avery (1988) discussed the role of ammonia in the hydrogen economy. He described the OTEC plantship concept for production of ammonia (George and Richards, 1980), which can be transported to the land using well established infrastructures. The JHU/APL OTEC plantship had nominal power generating capacity of 40 MWe, which is used for ammonia production of 104 metric tons per day (MTD). This plantship was intended to be a pilot plant for testing different power system components. A detailed design was performed and estimate of total deployment cost was prepared. The total deployment costs (in 1980\$), including the electrolyzer and ammonia plant, were \$177.2M (\$4,860/kW). Figures 3 and 4 show a schematic view of the plantship and the equipment layout, respectively.

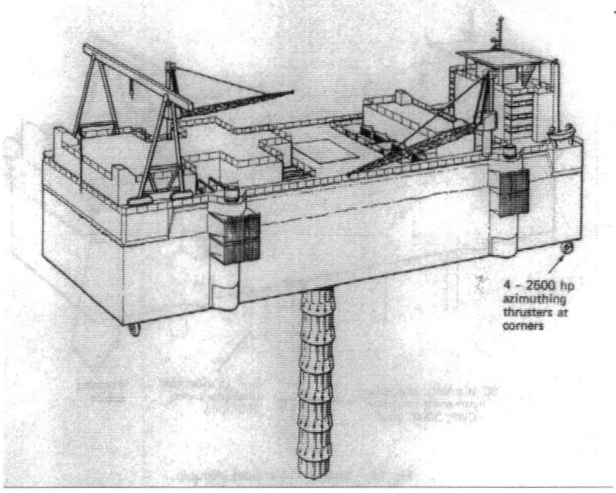

Figure 3. Schematic view of the JHU/APL OTEC plantship

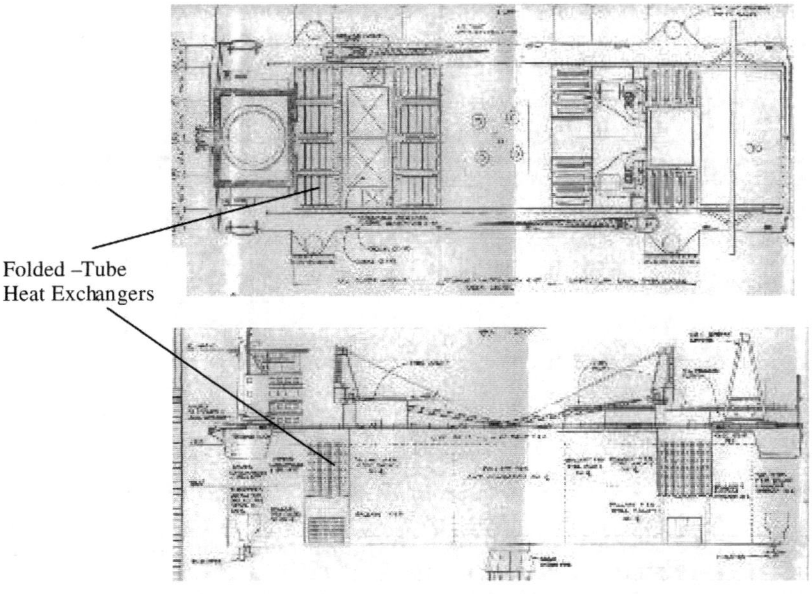

Figure 4. Layout of the JHU/APL OTEC plantship

The original OTEC plantship design incorporated folded-tube heat exchangers (Figure 5). They were cheaper to build; however, they occupied significant space on the plantship. Argonne National Laboratory, jointly with an industrial partner, developed compact high-performance brazed plate heat exchangers (Figure 6) based on commercial manufacturing processes. The thermal performances of prototype of both heat exchangers were determined at the Argonne's heat exchanger test facility. By keeping the same platform design and seawater system as the 1980 JHU/APL plant-ship design, and installing these compact brazed heat exchangers, the ammonia production capacity of the plantship can be increased from 104 MTD to 261 MTD for a 1.1°C less temperature difference than the 22.2°C used in the original design. The temperature difference of 21.1°C is representative of the Gulf of Mexico. The major design parameters of the OTEC power system are presented in Table 1.

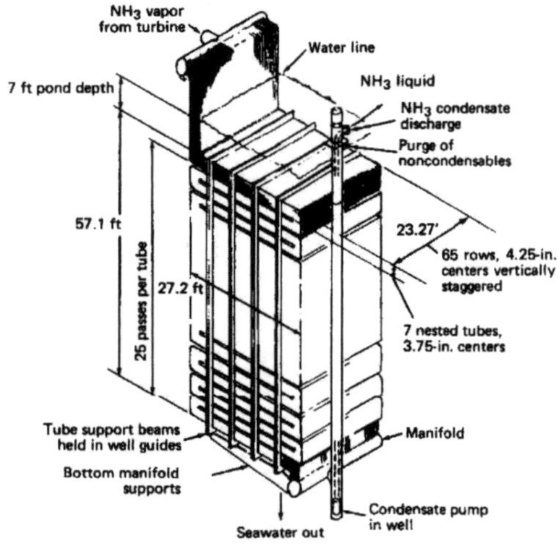

Figure 5. Folded-tube heat exchanger

Figure 6. Aluminum plate-fin heat exchanger

Table 1. Process parameters of the OTEC plantship

System Parameters	Ammonia Production Only	Co-Production of Fresh Water
Warm water temperature, °C	26.7	
Cold water temperature, °C	5.6	
Warm water flow rate, kg/s	325,970	413,850
Cold water flow rate, kg/s	275,130	364,140
Gross electric generation, MWe	129.6	129.6
Net power to ammonia plant, MWe	92.6	84.3
Hydrogen production, kg/hr	1,929	1,756
Ammonia production, metric ton/day	263	239.0
Fresh water production, million liter/day		114.0

The original JHU/APL plantship was not designed for production of fresh water. Therefore, to evaluate the economics of co-production of fresh water on ammonia costs, a hybrid-cycle plantship was analyzed. There are two hybrid-cycle concepts for water production. An integrated hybrid-cycle is more efficient with lower installed costs (Panchal and Bell, 1987). The add-on hybrid-cycle uses the conventional multi-stage flashing (MSF) technology (Rabas, Panchal and Genens, 1990). Although the add-on hybrid-cycle is not necessarily an optimum design, it was considered in this analysis to keep the cost estimates consistent with the original design of the JHU/APL plantship, and cost estimates are available for the MSF technology. The major design parameters for 114 million liters per day of water production are presented in Table 1. This is about one-half the water production capacity of the integrated hybrid cycle. The add-on hybrid cycle requires parallel seawater flows, so warm- and cold-water flow rates are increased by 27% and 33%, respectively. This results in increased pumping power, thereby reducing the net power to 82 MWe and the ammonia production to 239 MTD.

Economic Analysis

The Work Breakdown Structure (WBS) and total-installed costs for the first one-of-a-kind plantship and 10th plantship, with and without co-production of water, are presented in Table 2. The cost estimates of the platform, seawater system, and power system (except heat exchangers), and deployment were derived from the original design by applying *Chemical Engineering* cost indices from 1980$ to 2005$. Costs of brazed plate-fin heat exchangers were determined from the cost algorithm developed by Argonne. The cost of the water production subsystem was estimated from reported costs of MSF evaporators. The cost and performance of the electrolyzer was obtained from the reported data of Norsk Hydro system (Shinnar and Citro, 2007). The cost of ammonia synthesis plant with pure hydrogen as the feedstock is stated to be about one-third the cost of NG-based ammonia plant. The total installed cost of the NG-based ammonia plant was obtained from a reported cost of a plant in Trinidad ($268 for a turnkey construction of 1,850 MTD plant).

Table 2. WBS and cost estimates (2005$) of the OTEC plantship

WBS		First Plantship $ million	10th Plantship $ million	10th Plantship with water production $ million
1.1	Platform System	$99.39	$69.57	$91.83
1.2	Cold-water pipe system	$12.58	$8.80	$10.40
1.3	Power system	$154.58	$108.32	$108.32
1.4	Hydrogen electrolyzer	$55.24	$46.95	$46.95
1.5	Ammonia synthesis plant	$37.21	$26.04	$26.04
1.6	Water production plant			$62.62
1.7	Acceptance testing	$1.19	$0.83	$0.83
1.8	Deployment services	$17.61	$12.33	$12.33
1.9	Industrial facilities	$2.35	$1.68	$1.68
	Total engineering equipment	$380.15	$274.53	$360.97
2.1	Engineering and detail design	$5.03	$5.03	$5.03
2.2	Contingency	$38.02	$27.44	$35.80
	Total capital costs	$423.20	$307.00	$401.8

The capital costs as per the WBS shown in Table 2 were incorporated into the ***H2A*** Excel-based hydrogen economic program. The basic economic assumptions in the ***H2A*** program were used for this analysis. However, applicability of these assumptions to OTEC plantships will be further examined. The ***H2A*** program derives the cost of hydrogen ($/kg) for a given Internal Rate of Return (IRR) of 10%, which was transposed to ammonia costs ($/MT) as shown in Table 3. The table also shows the cost contributions from capital recovery, operation & maintenance (O&M), and credits for fresh water and oxygen. The nominal production costs of desalinated water with shore-based thermal and RO processes are about $\$1.3/m^3$ and $\$1.1/m^3$, respectively.

Table 3. H2A results of ammonia production costs (2005$)

	First Plantship $ million	10th Plantship $ million	10th Plantship with water production $ million
Hydrogen cost @ 10% IRR, $/kg	$6.11	$4.48	$2.04
Cost contributions Capital Operating & Maintenance Credit for fresh water & oxygen	 $4.71 $1.48 $0.08	 $3.42 $1.14 $0.08	 $4.32 $1.40 $3.68
Ammonia cost, $/MT	$1,078	$790	$360

The ***H2A*** results show the manufacturing cost of ammonia for the first one-of-a-kind plantship based on JHU/APL design will be significantly greater than the current market value of ammonia. The production cost of ammonia reduces to a lower value for the 10th plant with maturity of the technology; however, costs are still high. By co-producing fresh water, the cost of ammonia is significantly reduced. This analysis did not include the renewable energy credit, which should further improve the economics of OTEC plantships.

Conclusions

This design analysis presents the first part of an on-going project to evaluate the technical and economic viability of at-sea OTEC Plantships to produce ammonia, as the hydrogen carrier, and fresh water. The major conclusions from this first part of the analysis are as follows:

1. The ***H2A*** results show that OTEC ammonia plantship based on the JHU/APL design is a viable option for at-sea production of ammonia, as the hydrogen carrier, from non-carbon renewable energy sources by incorporating compact plate-fin heat exchangers and with co-production of fresh water; and
2. With further development of technologies and optimized design, such as an integrated hybrid-cycle and the new platform design, the economics can be further improved.

The project continues with further analyses, as shown below:

1. A new platform design based on the technology of the floating runway developed for DARPA;

2. A design analysis of the integrated hybrid-cycle OTEC plantships on the new platform;
3. A concept of a cluster of ten co-located 100 MWe OTEC plantships, which can feed electric power to an ammonia synthesis plant with a capacity of 2,500 MTD; and
4. An analysis of the logistics of transporting ammonia and fresh water from open seas to the US coastlines, specifically the Gulf of Mexico coastal states and California.

Acknowledgement

This work supported by the U.S. department of Energy, Assistant Secretary of Energy Efficiency and Renewable Energy, and the Office of Hydrogen Fuel Cell and Infrastructure Program, under Contract W-31-109-Eng-38. The authors acknowledge the support of Mark Paster, the DOE Program Manager of Hydrogen from Renewable Sources.

References

Avery WH, "A role for ammonia in the hydrogen economy," *Int J of Hydrogen Energy*, Vol. 13, No. 12, pp. 761-773, 1988.

Markey N, "Getting off the natural gas habit – Coal to nitrogen fertilizers," Presentation at the Gasification Technologies, Tampa, FL, March 2006.

Duijm NJ, Markert F, and Paulsen JL, "Safety assessment of ammonia as a transport fuel," *Risø National Laboratory Report*, Risø-R-1504 (EN), Denmark, February 2005.

George JF and Richards D, "Baseline design of moored and grazing 40-MW OTEC pilot plants, *The John Hopkins University/Applied Physics Laboratory Report*, JHU/APL SR-80-1A & 1B, June 1980.

Hacker V and Kordesch K, "Ammonia cracker," in Handbook of *Fuel Cells – Fundamentals, Technology and Applications*, editors Vielstich, Lamm and Gasteinger, John Wiely & Sons Ltd publisher, Chichester, UK 2003.

Ivy J, "Summary of electrolyte hydrogen production," *National Renewable Energy Laboratory Report*, NREL /MP-560-36734, September 2004.

Kubic WL and Belooussov AV, "Analysis of nuclear-powered ammonia plant," *Proceedings of the AIChE National Meeting*, Orlando, April 2006.

Panchal CB and Bell KJ, "Simultaneous production of desalinated water and power using a hybrid-cycle OTEC plant," *J of Solar Energy Eng.*, Vol. 109, 1987.

Rabas T, Panchal CB, and Genens L, "Design and cost of near-term OTEC plants for the production of desalinated water and power," *Solar Energy in 1990's*, ASME Publication SED-vol. 1, 1990.

Shinnar R and Citro F, "Upfront, efficient technology evaluation," Chem Eng. Prog., February 2007.

Stevens HC, Genens L, Panchal CB, "Conceptual design of a 10 MW shore-based OTEC plant," *ASME Paper*, 84-WA/Sol-31, 1984.

Thomas G and Parks G, "Potential role of ammonia in hydrogen economy," *DOE Report Freedom Car and Fuel Partnership*, February 2006.

Vega L, "Economics of ocean thermal energy conversion (OTEC)," Chapter 7 in *Ocean Energy Recovery" The State of the Art",* ASCE Publication, Reston, VA 1992.

P.E.M. Water Electrolyzers in Renewable Energy Capture to Hydrogen

Larry Moulthrop**, Jesse Hayes, and Mike Spaner
Distributed Energy Systems
10 Technology Drive
Wallingford, CT 60439

** To whom correspondence should be sent

Key Words: hydrogen generation, water electrolysis, renewable energy, proton exchange membrane

Prepared for Presentation at the 2007 Spring Meeting, San Antonio, TX, April 23-27, 2006

P.E.M. Water Electrolyzers in Renewable Energy Capture to Hydrogen

Larry Moulthrop, Jesse Hayes, and Mike Spaner

Distributed Energy Systems, 10 Technology Drive, Wallingford, CT, USA 60439

ABSTRACT

Distributed photovoltaic arrays and wind generators, enabled by distributed water electrolysis hydrogen generation and energy recovery fuel cells, all produced on a scale commensurate with today's automotive industry, may very well be the most compelling way to fill a growing world "carbon-free" energy gap. Hydrogen by electrolysis from renewable energy (RE) sources such as photovoltaic (PV) and wind has great potential to complement and extend the cost-competitive position and viability of RE against conventional fossil fuels and hydrogen production via steam methane reformation or coal gasification production methods, while avoiding emissions of greenhouse gases. High pressure (HP) proton exchange membrane (PEM) electrolyte water electrolysis is especially conducive to distributed hydrogen generation options where the cost of transporting hydrogen can be prohibitive. PEM water electrolyzers scale from 1 to 500 kilograms per day capacity, and are environmentally benign, using only pure water as feedstock and working fluid, with no liquid electrolyte. Commercial hydrogen generators using PEM water electrolysis are well-proven and are currently serving industrial and energy applications worldwide in over 50 countries.

Market and environmental requirements are converging to demand larger on-site hydrogen generators. Rising natural gas pricing, increasing trucking costs, developing economies with no liquid H2 infrastructure, utility load leveling, telecommunication power back-up, and forklift fuel cell fueling applications are all examples of market forces working to increase demand for on-site H2 generation. Intense international debate on global warming trends and the link to greenhouse gases from fossil fuels consistently cites energy from renewable sources as a 'must-do' pathway, yet the intermittent nature of solar and wind is often cited as a key drawback.

PEM Water electrolysis is the best enabler to intermittent renewable sources such as PV and wind. It can be dispatched quickly to convert RE as it is available to gaseous hydrogen at pressure; this stored or pipeline-supplied hydrogen may then be converted at-will back to energy via fuel cell or engine when needed.

INTRODUCTION

By the year 2050, a 17 Terawatt (17,000,000,000,000 watt) world "carbon-free energy gap" has been projected[1]. One way to fill this gap would be to commission a 1GWe nuclear power plant every day from now on, about 17000 additional 1GWe power plants by 2050. However, this seems to be a daunting task in light of environmental, security, and geopolitical issues associated with nuclear power plants; even today, there are only several hundreds of nuclear plants world-wide.

Now let's consider renewable energy, and solar PV as an example. There is about 600 TW of solar energy available world-wide, an excess well beyond our projected needs. Admittedly, today's production costs for solar PV is not economically viable at $6-8/watt installed. The huge challenge is to scale production to such a rate and capacity that the price falls to within $1-2/watt installed, which is economically viable, perhaps to the needed rate of 1GWe/day capacity. How could we get there? Dr. Robert Shaw of Arete Ventures[2] draws an intriguing analogy to the US auto industry which produces more than 15 million vehicles each year; each car has a 50kW power plant. That equates to 2GWe capacity produced each day, more than twice the projected 1 GWe need. So, Dr. Shaw postulates that an industry the size of one half of the US auto industry could potentially produce enough solar PV systems to meet the world's energy needs to fill the energy gap.

Renewables like PV and wind are intermittent in nature; some energy storage is necessary so that power can continue when the wind dies down or when the sky gets dark. Using renewable energy to produce hydrogen is perhaps the most logical option. Hydrogen readily replaces carbon-based fuels for transportation and can also generate electricity via fuel cells or combustion processes when there is no wind or sunlight.

A product development pathway has evolved at Distributed Energy for hydrogen generators in consideration of future needs. Vehicle fueling applications in the next 5-10 years may be well-satisfied with a 100 kg H_2/day PEM hydrogen generator module; this platform is then the pathway towards the 500 kg H_2/day module desired for hydrogen vehicle fueling stations, utility load-leveling, and renewables to hydrogen enablement. HP PEM units for reserve or backup power may be considerably smaller, as may individual vehicle fueling and home energy units.

Certain PEM electrolyzers generate hydrogen at high pressures while generating oxygen at low pressures. Elevated pressures of the hydrogen product are attractive for direct storage and help realize increased system efficiency and reliability. This reduces cost, footprint, components, and consumed energy compared to systems employing mechanical compressors.

Renewable energy is recognized for its low environmental impact when compared to hydrocarbon based energy carriers but fluctuations in its energy output are a technical hurdle. Each type of RE has an associated natural cycle or variability spanning minutes through decades. Geographic concentrations of RE and capital costs also vary greatly, highlighting the importance of developing a strategy for tapping these resources.

Generating Hydrogen and Oxygen from Water

Combining the environmental benefits of renewably generated electricity with the flexibility of PEM electrolysis, hydrogen can be generated to function as a clean energy carrier. Such a system has the following fundamental inputs and outputs:

Inputs

- **Water**: The electrolyzer system may include water processing capabilities.
- **Electricity**: DC current (renewable or grid) is required for the electrolysis process.

Outputs

- **Hydrogen gas** (H_2): The energy content of hydrogen is 39.4 kWh/ kg (higher heating value). Actual energy needed to generate hydrogen is dependent on the conversion efficiency of the electrolyzer system.
- **Oxygen gas** (O_2): The oxygen may be vented or utilized for other processes. Oxygen re-use may improve the economics of installing the electrolyzer.
- **Heat**: Heat is generated as a result of system inefficiencies. Larger electrolyzers may utilizing waste heat in combined heat & power systems.

- **Water vapor** (H_2O): PEM electrolyzers generate very pure hydrogen; the only impurity is water vapor. As electrolysis pressure increases, is becomes easier to manage the water content of the resultant hydrogen gas within specified limits.

The conversion from liquid water to hydrogen and oxygen is accomplished via one or more PEM electrolyzer stacks. Each stack consists of one or more electrochemical cells. The electrolyzer cell consists of anode and a cathode electrodes bridged by a bubble-tight solid polymer electrolyte membrane. Figure 1 illustrates the PEM cell reactions: the anode water electrolysis/oxygen evolution half-reaction, the proton exchange through the membrane, cathode hydrogen evolution half-reaction and the supply of electron current and water that sustains the reaction.

Anode: $H_2O \rightarrow O_2 + 2H^+$

Cathode: $2e^- + 2H^+ \rightarrow H_2$

Overall: $H_2O \rightarrow \frac{1}{2} O_2 + H_2$

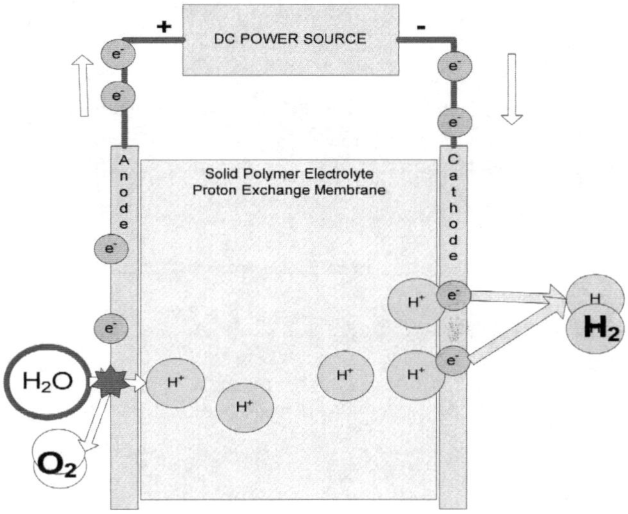

Figure 1

Simplified PEM electrolysis process

How High Pressure Hydrogen Generation is Accomplished

The hydrogen output volume from the electrolyzer is confined by a back pressure regulator or other device; the H_2 pressure builds until it exceeds the regulator spring setting. In this way, the H_2 is efficiently compressed by electrochemical means. The anode flow and output volume do not have to be confined in a high differential pressure PEM electrolyzer stack. A high differential pressure is mechanically sustained by the cells from anode to cathode. A simplified block diagram is shown in Figure 2. Generation of H_2 at elevated pressure requires more energy. The increase in cell voltage with respect to pressure is characterized ideally by the Nernst equation:

$$\Delta V = \frac{RT}{nF} \ln \frac{P2}{P1}$$

(Equation 1)

Where:

ΔV: *Cell voltage increase*

R: *universal gas constant, 8.314 J/mol, K*

T: *Temperature, K*

n: *number of electrons, 2*

F: *Faraday's constant, 96,485 C / electron-mol*

P_2/P_1: *pressure increase ratio*

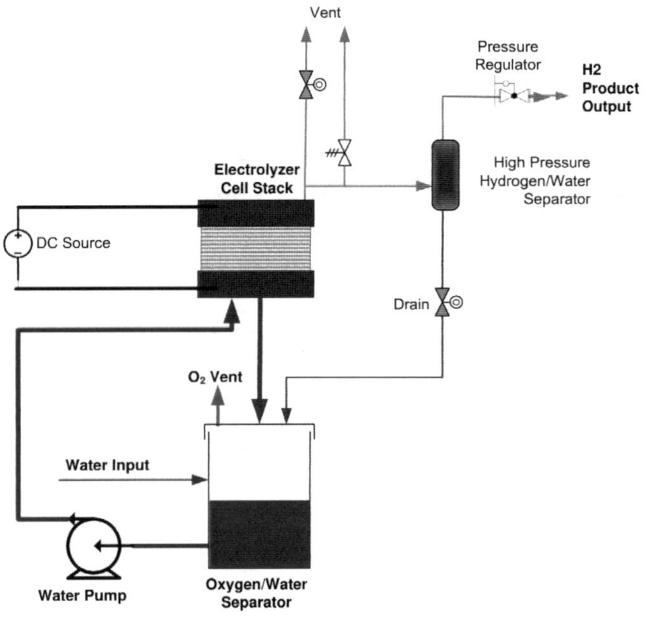

Figure 2
High Differential Pressure Electrolyzer Diagram

Actual cell voltages deviate from the ideal predicted by the Nernst equation due to factors such as changes in current density due to kinetic electrode effects, changes in mass transport, and increased hydrogen permeation. Back diffusion of hydrogen from the cathode to anode results in reduced hydrogen yield as pressure increases and input power is held constant. Figure 3 illustrates the rise in cell potential with generation pressure for a cell stack with 2000 hours of operation.

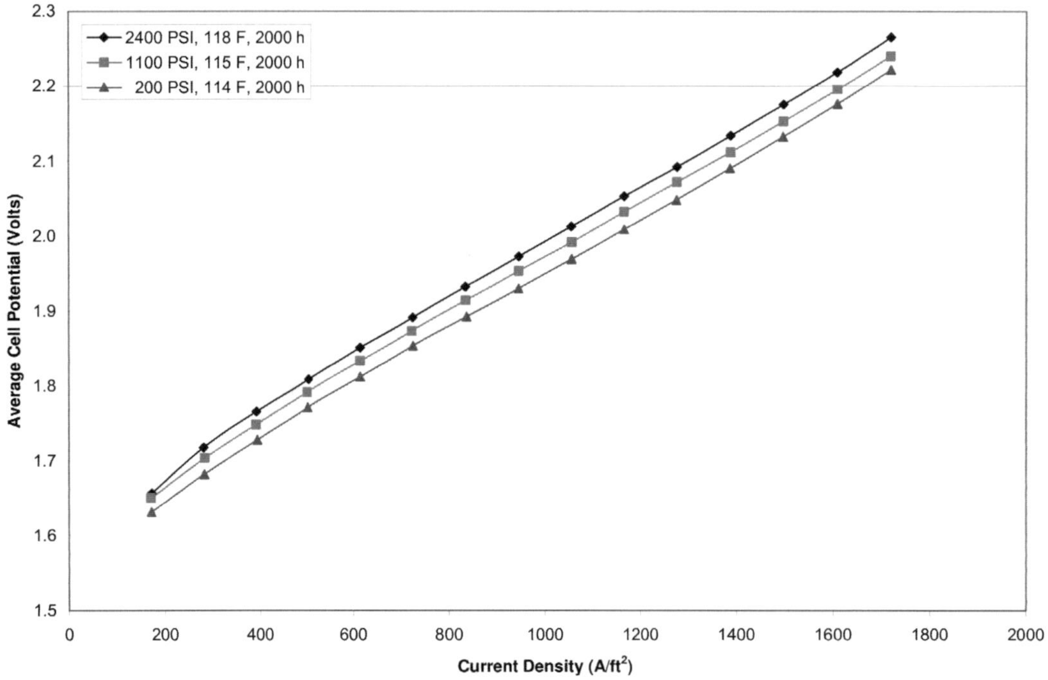

Figure 3 *High Differential Pressure Electrolyzer Polarization Curves*

High pressure generation and storage has significant advantages for distributed, renewable energy based systems. Consider the following scenario:

- 10 kW Fuel Cell System: consumes 70 SLPM hydrogen per 5 kW net output power
- Desired Energy Storage: 10 kW for 10 Hours (100kWH)
- HP300 Welding Cylinder storage: 49 Liters (water volume, each)
- Minimum Working Pressure required: 4.8 barg (70 psig)

- If the storage pressure = 27.6barg (400psig), tanks required = ~66 tanks
- If the storage pressure = 165.5barg (2400psig), tanks required = ~10 tanks.

Balanced high pressure electrolyzer systems have been developed by Distributed Energy Systems and others exceeding 200barg (2900psig)[3]. Incoming water and resultant oxygen may be compressed to eliminate high pressure differential pressure across the electrolysis cells. However, high pressure oxygen can be very dangerous. Special design and fabrication materials and methods are required to ensure appropriate levels of safety. Due to significant cost and complexity implications, balanced pressure systems have been most appropriate to niches such as military and aerospace applications, where high pressure oxygen is desired.

Storage Opportunities for High Differential Pressure PEM Electrolyzers

When hydrogen is stored as a gas in storage tanks, volumetric energy density is proportional to the pressure at which the hydrogen is stored. As storage pressure increases, the number of tanks, storage volume, and storage footprint decreases. This improved energy density is offset by the increased storage tank strengths required. Steel tanks are mature technology for stationary applications. Carbon fiber hydrogen storage tanks are more costly but are lighter weight for use in transportation applications. Metal Hydride hydrogen storage systems offer much higher volumetric energy storage densities, lower operating pressures, and easier handling and siting requirements. Drawbacks to metal hydride systems are low gravimetric energy density, system complexity, and cost.

Hydrogen stored at high pressures using PEM electrolyzers is achieved by:

1. Generating hydrogen at low pressure and using one or more mechanical compression stages to increase the storage pressure.
2. Electrochemically compressing the hydrogen to the storage pressure directly within the electrolyzer system. The complexities of mechanical compression are eliminated. Noise signatures are attenuated significantly. Costs may be reduced and system efficiencies improved with electrochemical compression. Differential pressures exceeding 200barg (2900 psig) have been accomplished by Distributed Energy Systems
3. Implementing hybrid systems using electrochemical and mechanical compression stages. Costs are optimized by minimizing electrolyzer system costs and reducing the size and number of mechanical compression stages required. Distributed Energy Systems successfully deploys fueling stations using this hybrid approach for use with automotive and fork lift fueling.

PEM electrolyzers are commercially available in the 15 to 30 barg (210 to 435 psig) range which permit industrial gas users to generate hydrogen on site to replace delivered hydrogen solutions (see www.protonenergy.com). Hybrid systems using electrochemical and mechanical compression are also being deployed for hydrogen fueling stations. The National Hydrogen Association Fueling database reports 72 stations in the U.S. and Canada. See *http://www.hydrogenassociation.org/general/fuelingSearch.asp* for details.

REFERENCES

[1] "Energy and Transportation: Challenges for the Chemical Sciences in the 21st Century"; The National Academies Press (2003)

[2] Shaw, R. W (2006), "Hydrogen, Enabling the Solar Option", Presented at the National Hydrogen Association Renewables to Hydrogen Forum

[3] Moulthrop, L (2004), "High Pressure Hydrogen Generation by PEM Electrolysis," Presented at the National Hydrogen Association's 15th U.S. Hydrogen Conference

Economical Evaluation and Japanese Research and Development Status of Cellulose-Based Biomass Utilization System

Akira Yabe[1], Kinya Sakanishi[2], Tomoaki Minowa[2], Shinji Fujimoto[2], and Masanori Kuroda[3]. (1) National institute of Advanced Industrial Science and Technology (AIST), 2-2-2 Hiro-Suehiro, Kure, 737-0197, Japan, (2) Biomass Technology Research Center (BTRC), National Institute of Advanced Industrial Science and Technology (AIST), Chugoku Center, Kure, Japan, (3) Biomass Technology Research Center, AIST CHUGOKU, Hiroshima, XXXXXX, Japan

For producing the energy material from the cellulose-based biomass resources, several challenges have been proposed and performed in Japan. For producing liquid automobile fuel from several kinds of cellulose-based biomass resouces such as sawmill residues and rice straw, National Institute of Advanced Industrial Science and Technology (AIST) of Japan has established Biomass Technology Research Center (BTRC) in AIST CYUGOKU. Some characteristics and advantages of several Japanese challenges have been introduced and the importance of estimating the system efficiency and economical analysis of system has been stressed for promoting the cellulose-based biomass utilization for preventing the global warming and decreasing the fossil fuel usage. As the actual analysis of the economical evaluation and a first step towards analyzing the production of automobile fuels from the cellulose-based biomass, the ability of wooden biomass energy conversion system to produce energy material has been analyzed on the basis of the material and energy balances. A zero- emission type biomass energy conversion system was analyzed to produce ethanol, lactic acid, and propanediol from cellulose and it supplies electric power and steam to satisfy the internal requirements of the system (wood throughput capacity: 420 ton/day). Payback periods were calculated from the raw material costs of wooden biomass and the major equipment cost of the energy plant; it was found that the payback period was within 9 years. Economic analysis indicated that sawmill residue and building waste wood, which cost less than $100/ton, were feasible in the proposed system.

AIChE Annual Spring Meeting
Houston, Texas
April 22-26, 2007

THE CANADIAN UNIVERSITY BIOREFINING NETWORK: AN R&D POWER HOUSE

Thomas Browne[1], Richard Kerekes[2,] and Jean Paris[3]
[1]. Paprican, Pointe-Claire, Québec
[2]. University of British Columbia, Vancouver, BC
[3]. École Polytechnique, Montréal, Québec

EXTENDED ABSTRACT

The pulp and paper (P&P) industry is a key component of Canada's national economy. In recent years, however, several factors have created a very difficult economic environment for this industry and it must make radical changes to return to profitability. Biorefining must be part of any strategy for change. This term designates the utilization of forestry and agricultural biomass to produce a spectrum of products by various extraction and transformation pathways. Chemical pulping processes are prime candidates for conversion to biorefining. The implementation of biorefineries in Canada by converting a substantial segment of its pulping production capacity in a timely fashion will require the deployment of massive efforts covering a broad range of activities.

PAPIER and Paprican formed a joint Task Force to plan, promote and coordinate a Canadian initiative on forest biorefining with particular emphasis on the role and place of Canadian universities and research institutes in this undertaking. PAPIER, the Canadian Pulp and Paper Network for Innovation in Education and Research is an association of Canadian university faculty involved in research for the P&P sector. Paprican, the Pulp and Paper Research Institute of Canada, is a world-class industry-supported research institute.

The Task Force has proposed to key stakeholders a plan for the creation of a Canadian network for the implementation of a forest biorefining industry; it is the Canadian Forest Biorefining Network (CFBN). This network would be very board in scope, encompassing short-term to long-term projects, university-lead research and development as well as industry-lead large-scale demonstration projects.

The Task Force has also compiled and defined 12 research, development and demonstration domains in which work will be required. The list of the 12 domains is given below in order of priority. The CFBN will be under the management of the new institute, FPInnovations, which has recently been formed by merging Paprican, FERIC and Forintek. Funding for the CFBN is currently under consideration by the management of FPInnovations.

RD&D Program Domains

1. Product opportunity analysis
2. Forest feedstocks
3. Thermochemical pathways
4. Bioproducts from effluent and solid wastes
5. Products from hemicellulose
6. Products from lignin
7. Products from extractives
8. Products from condensates
9. Phyto-chemical products from bark, branches and foliage
10. Integrating novel products with existing product lines
11. Conversion of uncompetitive pulp mills
12. Integrating pulp mills with other industries

An informal survey of university activities pertinent to biorefining conducted by PAPIER has shown that there is a large potential for high level scientific work in Canada that could be mobilized to contribute to the RD&D program envisaged. The response has been very enthusiastic and the current roster contains about 100 faculty members from 15 Canadian universities engaged in a broad spectrum of R&D activities.

A number of these activities are likely to become components of the CFBN. However the range of biorefinery activities undertaken in the academic community may be broader and more fundamentally oriented than in the CFBN. They will also have an important training function. For those reasons, the needs of the academic community will be different from those addressed by the CFBN. To address those needs PAPIER has decided to create an independent university biorefinery network called the Canadian University Forest Biorefinery Network, CUFBNet. This network will collaborate fully, whenever possible, with the CFBN and, it will be proposed that it become a component of the CFBN in some appropriate manner.

CUFBNet will be a voluntary associative network of university-based projects.

It will provide an organisational structure for information sharing, coordination and cooperation between research groups and, development of partnerships with research institutes and industry. It will also constitute a framework for joint grant applications.

CUFBNet does not have any operating budget at the present time; however, it is anticipated that adequate funding for its operation will be secured in a reasonably short term. The purpose of the budget will be to operational costs since CUFBNet is not intended to be a significant R&D granting agency.

The specific objectives of CUFBNet are:

- Provide a structure for information sharing, networking, developing collaborations and partnerships;
- Act as a common voice for the Canadian university forest biorefinery research community when needed;
- Interface with government and granting agencies, support and coordinate grant applications;
- Link with international network counterparts;
- Act as single-source access for government and industry of up-to-date information on university-based biorefinery R&D in Canada.
- Organize meetings of the network membership and other joint activities such as symposia and workshops.

Membership in CUFBNet will be open to all members of the Canadian academic community involved in forest biorefinery and related R&D. Benefits to members will be:

- Single-source access of up-to-date information on biorefinery activities among members and on development outside of Canada.
- Linkage to other R&D groups and projects, development of collaborations and partnerships, opportunities for joining existing or new projects, participating in grant applications.
- Single, recognised representation in CFBN and participating agencies.

University individual scientists and teams joining CUFBNet will be required to accept some obligations such as:

- Provide timely information not restricted by obligations to funding sources or as it enters the public domain;
- Participate in meetings and other CUFBNet sponsored activities;
- Join CUFBNet R&D projects when appropriate and as they may be formed.

As CUFBNet develops its own financial resources, it will offer active support for the preparation of project proposals and grant applications.

As a partner in the task-force, Paprican will also participate in CUFBNet; it will contribute information just entering the public domain or otherwise not restricted by obligations to financial supporters, it may provide letters of support to grant applications for projects that fit into its own strategic plan; it may actively participate in specific projects when appropriate and make in-kind contributions.

CUFBNet is currently engaged in the following short-term actions:

- Undertaking an intensified effort to secure adequate funding for the execution of its mandate and to achieve its objectives as stated above.
- Gathering detailed information from the members of the Canadian university community to identify potential networks.
- A website of all biorefinery academic research in Canada will be created as an information resource for government and industry.
- A plan of action to create project networks is being developed.

Process Analysis and Optimization for Transportation Liquid Fuel Production by BTL (Biomass to Liquids) Process

Shinji Fujimoto, Takashi Yanagida, Kenji Asano, Akira Yabe, and Tomoaki Minowa. Biomass Technology Research Center, National institute of Advanced Industrial Science and Technology (AIST), 2-2-2 Hiro-Suehiro, Kure, 737-0197, Japan

Biomass, that is carbon neutral and a renewable resource, becomes attractive after the decision of Biomass-Nippon strategy in Japan. A process to produce transportation liquid fuel from biomass is also attractive. In this study, BTL (Biomass to Liquids) process was considered from viewpoints of economics and environmental. The process comprises gasification of biomass, gas cleaning, compression of syngas, FT synthesis and distillation. That is, in the process, biomass is converted into syngas at first. Then syngas is cleaned and is pressurized. The pressurized syngas is synthesized into hydrocarbons. Finally, FT diesel is obtained by distillation. We are studying the economic and environmental evaluation based on process simulation and process analysis. The process requires thermal energy in gasification and distillation and electricity in compression. In this study, the effect of CO2 mitigation when the requirements were fulfilled by fossil fuel or biomass was considered. Consequently, the combustion of a part of a biomass as a heat source decreased the total CO2 emission though the partial combustion decreased the feedstock biomass and the yield of FT diesel. Further, the power generation using biomass was not useful if the efficiency was low. In other words, a process that does not depend on a fossil fuel is not necessarily effective. In the process, much amount of off-gas was discharged after FT synthesis. The off-gas had approximately a half of heating value of feedstock biomass. Therefore, the combustion of the off-gas and the recycle of the off-gas for effective utilization of the off-gas were examined. Consequently, it was found that the combustion of the off-gas as a heat source was effective for CO2 mitigation. On the other hand, the recycling of the off-gas to the gasifier where hydrocarbons were reformed into CO and H2 remarkably increased the yield of FT diesel. However, it increased power to compress the syngas and the consumption of electricity because of the increase of syngas flow rate. Accordingly, effect of CO2 mitigation decreased in this case. The compression capacity strongly affected CO2 mitigation. Furthermore, it was also strong factor in economics. That is, in the process, the cost of compressors accounted for 1/3 of the construction cost. In an improvement of the economics, the compressor section is sensitive. Therefore, a pressurized gasification that requires no compressor was examined and was compared with the atmospheric gasification. In this case, the cost of the pressurized gasifier increased because it had to be constructed heavily to endure high pressure. However, exclusion of compressors more strongly affected the economics than addition of the pressurized gasifier and the economics was improved.

Clean Gas from Biomass by Catalytic Filtration

Lina Ma and **Gino V. Baron**. Dept. Chemical Engineering, Vrije Universiteit Brussel, Pleinlaan 2, Brussel, B-1050, Belgium

Biomass gasification is regarded as a good candidate for localized, small scale renewable energy production, but will only be broadly implemented if simple and efficient technology is available for removing undesired impurities such as particles and tar from the producer gas in order to avoid serious operating problems. The same holds for larger scale gasification for the production of synthesis gas for conversion into fuels, chemicals or hydrogen. This work focuses on the further development and improvement of a novel one-step high temperature gas cleaning catalytic based on a candle filter for the simultaneous removal of tars and other impurities and particles. After a brief review of gasification gas cleaning issues, synthesis of the catalytic filter (several recipes), results on a laboratory setup, long term testing and results on a biomass gasifier are given.

An optimized nickel-based alumina ceramic catalytic filter material for the use in integrated high temperature removal of tars and particles from biomass gasification gas has been extensively tested in a broad range of parameters allowing the identification of the operational window of such a filter. The optimal composition of the material determined from screening a large number of combinations is a porous alfa-alumina filter material with pores coated by 2.5 wt% Al_2O_3, 1 wt% Ni and 0.5wt% MgO. At a typical face velocity of 2.5cm/s, in the presence of H_2S and at 900°C, the conversion of naphthalene in very high 20g/m3 tar loading is almost complete and a 1000 fold reduction in tar content is obtained. Technically, it would be better to run the filter close to the exit temperature of the gasifier around 800 - 850°C. At 830°C, conversions of 99.0% could be achieved in good gasifier operation conditions with typically only 2g/m3 tars, but further enhancement of the performance at low temperatures is desirable.

Results of tests of several types of catalytic filter (Ni based with different supports, modifiers and dopants) were collected with a particle free synthetic gasification gas with 50 vol% N_2, 12 vol% CO, 10 vol% H_2, 11 vol% CO_2, 12 vol% H_2O, 5 vol% CH_4, 0-200 ppm H_2S, and the selected model tar compounds: naphthalene and benzene. Both performance data in a wide range of temperatures, gas composition, tar loading, flow rate were obtained and a simple kinetic model for the conversion derived.

Furthermore, a method was developed to deposit a new support ZrO_2 material to increase the surface area of the alumina candle filter and further enhance the catalytic activity, especially at lower temperature. A series of filter discs were prepared by this urea method in order to compare the catalytic performances of catalysts containing different types of supports: ZrO_2, Al_2O_3 and a mixed ZrO_2-Al_2O_3 over a broad range of parameters. It indicated that the tar conversion achieved with the three types of supports: ZrO_2, Al_2O_3 and a mixed ZrO_2-Al_2O_3 are similar, but at high velocity and low temperature, the best is the mixed support, which can probably be further optimized. The mixed material also exhibited a better stability (tests over hundreds of hours) compared to the pure alumina based recipe.

Although the developed candle preparation method has been proven to be scalable, alternative preparation techniques for large scale production will also be discussed.

The Co-Production of Ethanol and Electricity From Carbon-based Wastes

A report from Bioengineering Resources, Inc. regarding a New Technology that addresses multiple Energy and Waste Disposal solutions

March 2007

Bioengineering Resources, Inc.

THE CO-PRODUCTION OF ETHANOL & ELECTRICITY FROM CARBON-BASED WASTES

Carbon-based wastes represent one of the world's most promising and virtually untapped renewable energy sources. Every year, the United States generates approximately 1.5 billion tons of municipal solid waste, biosolids, animal wastes, green wastes, food production and pulp paper wastes, plastics, auto shredding residues, agricultural residues and forestry wastes--320 million tons of which are *readily available* for conversion to energy in our local communities.

During the past 25 years, government and private industry in the United States have spent some $9 billion attempting to develop economic and environmentally sound methods for the production of electricity and liquid energy from these resources.

Despite this effort, only about 3% of America's liquid fuel and electricity currently comes from biomass and even less is generated from organic waste. The United States as a whole gets about half of its electricity from coal. Coal-fired power plants account for more than one-third of its greenhouse gas emissions and represent the nation's largest single source of industrial air pollution.

According to the Department of Energy, the United States imported 66% of its crude oil and petroleum products in 2006, severely impacting the nation's balance of payments and its national security. Americans, the world's largest gasoline consumers, used an average of 390 million gallons a day. The price of oil reached $78 per barrel during the year and we are now spending nearly $1 billion per day to import petroleum.

Milton Copulos, President of the National Defense Council Foundation, estimates that the "real" cost of gasoline refined from Persian Gulf oil is $10.86 per gallon.

The world's reserve capacity has been soaked up by increased demand, and is now estimated to be about two million barrels per day--in a world that every day is using 85 million barrels (or nearly 3.6 billion gallons) of petroleum products.

Global energy demand is expected to increase 54% by 2025, during which time demand for electricity will almost double. Crude oil consumption will increase by 31% to 111 million in 2025, when fossil fuels are expected to reach 85% of the world's primary energy mix.

America's oil consumption is projected to increase 32% and China's is expected to increase by 119%. By 2025, China will be using about half as much petroleum as the United States.

According to the New York Times, "China will surpass the United States as the world's largest emitter of carbon dioxide by 2009, a decade ahead of previous predictions. A big reason is the explosion in the number of automobiles, but the main reason is China's ravenous appetite for coal, the dirtiest of all the fuels used to produce electricity. Already, China uses more coal than the United States, the European Union and Japan combined. Every week to 10 days, another coal-fired power plant opens somewhere in China, with enough capacity to serve all the households in Dallas or San Diego."

The North American Electric Reliability Council says that by 2015 the United States will need an additional 141,000 megawatts to accommodate an expected 19% increase in electricity usage. Only 57,000 megawatts are on the drawing board.

The National Academy of Sciences predicts that hydrogen-powered vehicles won't be readily available for another 25 years, and before that time, the Natural Resources Defense Council estimates that Americans will buy 450 million new cars and trucks.

Increased domestic oil production cannot possibly keep pace with the nation's growing demand for liquid energy, and in this competitive environment, with petroleum becoming more difficult to access and bring to market, inflationary pressures on the price of crude oil are destined to continue.

We live in a petroleum-based economy and this is not going to change any time in the near future. Therefore, America's only solution is to blend an environmentally safe volume extender with gasoline to reduce the nation's dependence on foreign oil and its cost of liquid energy. Only one substance can achieve this and that is ethanol—and ethanol is already integrated into America's liquid energy distribution system.

Ethanol is now being used around the world in gasoline blends ranging from 5.6% to 85%. (For example, 40% of all new cars being sold in Brazil are capable of burning 100% ethanol and all gasoline sold there contains at least 25% ethanol. Brazil will soon be energy independent.)

Ethanol improves octane performance. Further, Michael Wang of the Argonne National Laboratory reports that cellulosic ethanol will achieve greenhouse gas (CO_2) emission reductions of 85% per gallon of ethanol used to displace an energy-equivalent amount of gasoline, even with only 10% blending.

Ethanol, to date, has been produced chemically from ethylene or biologically from the fermentation of sugars from carbohydrates found in agricultural crops like corn kernels. Thus far, sugar fermentation has been the only process to commercially produce ethanol from biomass. Historically, the industry has relied on state and federal subsidies, including a 51-cent per gallon federal subsidy and additional incentives ranging from five to 28-cents per gallon in the corn-producing states.

Currently, 110 American ethanol refineries have the capacity to produce 5.4 billion gallons of ethanol annually. Another 73 ethanol refineries and 8 expansions are under construction, more than a 50% increase in construction activity since mid-2006. These plants will yield an additional 6.0 billion gallons in annual ethanol production by 2008, exceeding by almost four billion gallons the 2012 Renewable Fuel Standard mandated by Congress in 2005.

(According to the USDA, 55 million tons of corn, or one fifth of America's 268-million-ton harvest, were used to produce ethanol in 2006. New plant construction has already taken 2007 corn futures above $4.00 per bushel, a trend that is impacting the price of corn in livestock farming, food and exports, and which is placing the profitability of corn ethanol production at risk.)

However, the nation is just taking its first steps down the road to energy independence. Iraq, Saudi Arabia, Venezuela and Nigeria—all of which are considered to be politically vulnerable—provide 38.2% of America's oil imports and 22.6% of its total consumption.

To replace America's petroleum imports from these four nations would require 32 billion gallons.

True energy independence, meaning generating all of our liquid energy from domestic sources, would require the nation to domestically produce the equivalent of 91 billion gallons of gasoline from sources other than petroleum.

If electricity and ethanol (and in the future, hydrogen) could be produced profitably from America's waste streams, it could help to stabilize the nation's supply of electrical energy, supplement or replace gasoline, convert vast quantities of waste to energy while reducing communities' costs of waste disposal, reduce greenhouse gases and significantly reduce the nation's dependence on foreign oil.

The BRI Renewable Energy Process makes these goals achievable.

The BRI Renewable Energy Process

During 2007, Bioengineering Resources, of Fayetteville, Arkansas, will commence construction on the world's first commercial plants that will co-produce low-cost ethanol and electricity from any carbon-based materials, including:

- Municipal solid waste
- Biosolids (sewage sludge) and animal wastes
- Green waste
- Corn stover and other agricultural residues
- Timber and wood wastes
- Used tires and plastics
- Auto shredding residues
- Pulp paper, food processing and other industrial wastes
- Coal, petroleum coke, natural gas and other hydrocarbons
- Methane (landfill gas)
- Stranded gas
- Refinery Tars & Waste Oils

Developed by a team led by Dr. James L. Gaddy of Fayetteville, Arkansas, this new gasification/biocatalytic process utilizes an enzyme from a patented bacterial culture that ingests synthesis gases (thermally decomposed wastes) and reconstructs them into fuel-grade ethanol. The process yields up to 85 gallons per dry, ash-free ton of biomass. From high BTU content materials like plastics, used tires or hydrocarbons, it can yield 150 gallons or more per ton.

The BRI technology is a ***generation beyond*** cellulosic ethanol, which is defined as the production of ethanol from purpose-grown plant materials, and which most experts believe is still 5-7 years away. The major difference is that, whenever "cellulosic" technologies are perfected, they will convert only the 50% cellulosic portion of plant materials into sugars and then to ethanol. The BRI process will always be almost twice as efficient, as it also converts the hemi-cellulose and lignin, or in total, 95-98% of plant materials.

The BRI process will gasify any carbon-based material whose moisture content is less than 30% (by weight). Its **feedstocks** can be blended and need not be chipped, shredded or sorted. Any mixture of plastics, tires, manure, paper or yard wastes, construction debris, furniture, crop residues, timber slash, etc., can be converted into synthesis gas, and then to ethanol.

Before being introduced to the bacterial culture in the bioreactor, the synthesis gases (CO, H_2 and CO_2) are scrubbed, put through active carbon filtration and cooled to approximately 98°F—a process that generates an enormous amount of waste heat that can be used to create high temperature steam to drive electric turbines. This can supply up to 50M BTU per gallon of ethanol produced. Gasification eliminates the need to combust coal in the generation of electricity, removing a major source of industrial pollution.

In the biocatalytic step that takes two minutes or less, the bacteria ingest the syngas and emit ethanol and water, which is then distilled away to produce 99.5% pure industrial or fuel-grade ethanol.

The entire process, from the time the waste material enters the gasifier to finished ethanol, takes about seven minutes, whereas corn ethanol fermentation requires 36-48 hours. As opposed to sugar fermentation technologies, the process is odorless and sterilization is unnecessary, since the carbon monoxide in the syngas eliminates most contaminants.

Sugar fermentation requires sterilization and batch processing. The BRI technology is continuous and consumes up to five times less water.. The bacteria will thrive and produce ethanol indefinitely--so long as fuel is entering the gasifier.

Rapid biochemical conversion, and the fact that the process creates up to five revenue sources, are features that make the BRI technology highly profitable and competitive with gasoline. Net of tipping fees and green power revenues, the process can produce ethanol for about one-third of the current retail cost of gasoline, even without federal or state subsidies.

As the process uses waste products that otherwise would have been placed in landfills and BRI's plants will generate an excess of electricity beyond their parasitic needs, they can produce liquid and electric energy while consuming zero new BTUs in the process. This makes the current discourse about the energy efficiency of ethanol obsolete.

The bacterial culture is anaerobic and dies when exposed to air. The process creates no environmental or health hazards, ground or water contamination, and minimal air emissions. When biomass is used to co-produce ethanol and electricity, significant reductions in greenhouse gas emissions can be achieved.

BRI's plants will also operate on natural gas, petroleum coke and coal--and these hydrocarbons can be blended with biomass to increase by up to 100% the overall gallon-per-ton output of a plant. Gasification eliminates the need to combust coal in the generation of electricity, removing a major source of industrial pollution. The technology can reform natural gas and landfill methane into ethanol.

The process will normally convert more than 80% of the waste it receives. The residue, a non-hazardous ash, is discharged from the gasifier and can be recycled in products like cement blocks or paving. The net effect is that the BRI process can extend by up to 80% the effective life of a landfill (and it can reclaim and convert into productive energy materials already residing in landfills).

If and when fuel cells become available to power automobiles, the BRI process can also be used to create hydrogen.

A Typical BRI Renewable Energy Plant

BRI's plants will be modular, and by adding modules, their capacities can be readily expanded. A single module will combine two complete energy production lines, each with a capacity to process 125 to 150 tons of waste per day. Each module will process up to 100,000 tons of biomass annually, and depending upon the BTU content of the feedstock, will produce approximately 6-8 million gallons of ethanol from green waste or MSW and generate 5-6 MW of power. The amount of ethanol and electricity to be produced by any module can be varied according to energy demand.

Among other configurations, a mid-sized BRI Renewable Energy Plant would employ seven modules to produce approximately 50 million gallons of ethanol and generate approximately 35 MW of power (about 21 MW of which would be available for sale into the grid). It would process 700,000 tons of municipal solid waste or other organic materials per year.

The combination of electrical generation and low-priced ethanol production (even if federal subsidies were to be phased out) makes possible long-term firm and stable contracts for the generation and sale of "green power" at between five and ten cents per kWh. On the ethanol side, assuming a conservative tipping fee for MSW of $18 per ton, the BRI plant will realize a negative feedstock cost of about $.25 per gallon, creating a significant production cost advantage over corn-based ethanol technologies.

Status of Technology

It has been 17 years since the first bacterial culture to convert synthesis gas into ethanol was isolated by Dr. James L. Gaddy and his technologists at Bioengineering Resources, Inc. After a long and patient technology development process, some 50 patents have now been awarded or are pending worldwide for the microorganisms, process and methods.

Since November 2003, when BRI added a prototype Consutech gasifier, the pilot plant has been operating the complete ethanol production cycle, from waste gasification through the delivery of ethanol, in a single integrated process. Approximately $15 million in investment, DOE grants and internally generated funds have supported the technology's development.

Katzen International, a Cincinnati-based engineering firm that is renowned for the efficiency of their ethanol separation and distillation technologies, has been closely involved with the BRI process during its entire pilot plant phase. Katzen has designed some 75 ethanol plants and currently has 20 ethanol-related projects on-going around the world. Their technology and expertise will be utilized to extract commercial grade ethanol from the fermentation tanks. Katzen will be responsible for the process design for BRI's plants.

In late 2005, Parsons Corporation, one of the world's leading engineering firms, delivered an independent Technology Evaluation Report and emissions study on the BRI process. Parsons will be updating its technology evaluation during the month of March, 2007. These reports will enable BRI to complete the permitting of its initial commercial plants. These plants will demonstrate the viability of such feedstocks as municipal solid waste, auto shredding residues, wood waste, sugar cane bagasse, purpose-grown (cellulosic) plant materials, coal gasification to ethanol and electricity, and the conversion of landfill methane to ethanol.

On February 28th, the Department of Energy announced that BRI and its joint venture partner had been selected for a $33 million grant to cover 40% of the construction cost of a cellulosic ethanol plant in LaBelle, Florida. This will help to build a 13.9 million gallon plant over two phases that will use wood wastes and purpose-grown plant materials to produce ethanol and electricity, and ultimately fertilizer.

The BRI Process & America's Waste Resources

Approximately 1.5 billion tons of municipal solid waste, green waste, sewage sludge, plastics, auto fluff, used tires, agricultural, forestry and other waste products are generated in the United States each year, 320 million tons of which are *readily available* for use in the production of liquid and electric energy.

And when organic waste materials are utilized as fuel, the technology supports the natural environmental cycle of CO_2 generation and recovery.

In addition, America has a three hundred year supply of coal, the combustion of which represents one of its most destructive sources of industrial pollution. Millions of dollars are being spent to develop coal gasification projects, but these technologies must still combust the resulting syngas to generate electricity. The BRI Renewable Energy Process can create electricity without combustion.

It is the only technology that can blend any of the resources listed above as feedstocks—and it is the only technology that can simultaneously co-produce renewable liquid energy, green power and other products from organic wastes.

Net of its revenues for tipping fees and green power, it can produce ethanol for blending with gasoline for less than one-third of the current average retail cost of gasoline and can deliver electricity (green power) in the range of 5-10-cents per kWh.

It could provide the nation's farmers and dairy industries with productive alternatives to the open-field burning of agricultural residues and the disposal of animal wastes.

It could extend by up to 80% the useful lives of existing landfills, while at the same time, lowering the cost of waste transport and disposal for municipalities. It can significantly reduce greenhouse gas emissions from the decomposition of landfilled materials, and can convert landfill methane gas into ethanol. It holds promise to eliminate the need for public agencies to spread sewage sludge on agricultural lands, usually in some other county or state than their own.

The BRI process will create a new revenue stream for farmers without having to change their crops. Corn growers, for example, will be able to generate income from their corn stover, as well as their corn. DOE-funded studies have demonstrated that major synergies can be achieved by collocating BRI renewable energy projects with traditional sugar fermentation plants. Farmers could deliver their entire corn plant to a single location for processing. The corn kernels would continue to be used as feedstock for sugar fermentation and the production of distilled wet and dried grain solubles for animal feed, and the remaining portion of the corn plant would be gasified by BRI to co-produce ethanol and such by-products as electricity, steam and ammonia for fertilizer. The sugar fermentation plant would have access to low-cost electricity and steam—and ethanol filtration, denaturation, storage, marketing and shipping, railroad and highway access, plant maintenance and administration facilities could be shared, lowering the cost of operation and increasing the profitability of both facilities. Energy costs for operation of the fermentation plant could be cut as much as 30%.

The technology could turn states like New York and California into major exporters, rather than importers, of ethanol. In California, for example, 40 million tons of post-recycled municipal solid wastes are placed in landfills each year, and this is expected to grow to near 50 million tons annually by 2025. Theoretically, the BRI technology could annually co-produce some 2.7 billion gallons of ethanol (approximately three times the amount that was imported to California from the Midwest last year) and some 2500 MW of green power--just from the state's post-recycled municipal waste streams--forgetting such other potential fuels as forest thinnings, agricultural residues, sewage sludge, animal wastes and manufacturing wastes. California must also dispose of 33 million used tires per year, one-third of which are put into landfills, and which, like plastics, do not degrade. Therefore, the technology offers the potential to “mine” and recover material from landfills and transform approximately 80% of those materials (the organic fraction) into ethanol and electricity, as well.

The BRI process, which is known as a conversion technology, offers one of the best avenues for electric utilities to meet their increasing mandates for the production of renewable electricity (green power), while reducing our dependence on coal-fired electrical power generation, the nation’s leading source of industrial pollution.

(Greenhouse gases rose in the United States by 2 percent in 2004, according to the U.S. Energy Information Administration. Along with carbon dioxide, methane and nitrous oxide, the nation produced 7.12 million metric tons of these gases, up from 6.98 million metric tons in 2003. That's 16 percent greater than in 1990, or about 1.1 percent average annual increases. Most of the increase comes from carbon dioxide and specifically, from burning such fossil fuels as coal, natural gas and oil.)

The technology can create a new industry and “green collar” jobs for America’s domestic work force, rather than exporting in recyclable materials to China, where there are no environmental regulations and where these materials are turned into products to be sold back into the United States.

The complete development of the Alaskan National Wildlife Reserve would provide only two percent of America’s liquid energy needs. It is possible to project that the BRI process could deliver 10% of America’s liquid energy requirements from locally generated waste streams within ten years.

The bottom line is that America has enough organic waste material and hydrocarbons to achieve energy independence.

The concept that today's waste streams can become tomorrow's liquid and electric energy supersedes all other solutions in our quest for energy independence. It is the paradigm for the 21st Century.

Contact:

James L. Stewart
Bioengineering Resources, Inc.
3142 Brookdale Road,
Studio City, California 91604-4207
323-650-5095; 818-261-2599 (mobile)
jls.bri@roadrunner.com

Production and Separation of Fermentation-Derived Acetic Acid

Seth W. Snyder, Energy Systems, Argonne National Laboratory, ESD-362, Argonne, IL 60439

Acetic acid is a large volume chemical product. Currently, 80 percent of the U.S. production of virgin acetic acid is produced by methanol carbonylation. Production costs are strongly tied to natural gas. Therefore, there are strong market, economic, and energy benefits to developing processes for production of acetic acid from alternative feedstocks. The advantages of producing acetic acid by fermentation include its appropriateness for regional-scale production, lower cost feedstocks, suitability for low energy membrane-based purification, and lower temperature and pressure requirements. In this project, we developed strains for fermenting syngas to acetate and evaluated several membrane-based separations technologies for recovery of the acid.

Web Page: www.eere.energy.gov/industry/chemicals/pdfs/acetic_acid.pdf

Energy Optimization Studies to Enhance Both New Projects and Return on Existing Assets

X. X. Frank Zhu, UOP LLC, 25 East Algonquin Road, Des Plaines, IL 60071

The demand for refined products has risen steadily in recent years at rates greater than both crude oil and refining capacity. As a result, the supply demand balance has tightened for both crude oil and refining capacity. These circumstances are driving several key trends in the refining industry. Energy prices have increased significantly in the US and throughout the world. Most projections call for energy prices to remain relatively high in the future. Also, investment and projects for new refining capacity have increased dramatically in the last few years. New and improved energy optimization techniques can enable refiners to address both of these trends. Energy optimization can enable refiners to significantly improve the economics of new capacity projects and also reduce energy costs in existing facilities at the same time. This is achieved via the combination of the latest pinch analyses and tools coupled with process knowledge and optimization. Case studies from recent projects will be used to illustrate how energy costs have been substantially reduced with very quick payouts not just in new capacity projects, but also in existing refining facilities. Several case studies will be presented for projects ranging from addition of single major units to refinery expansions involving several process units.

Decision Support Solution for the Operational Improvement of Refinery Pre-Heat Trains

Piyush Shah, Aspen Technology, 2500 CityWest Blvd, Houston, TX 77494, Hernan Rodera, Engineering & Innovation, Aspen Technology, 2500 CityWest Blvd, Houston, TX 77494, and Hiren Shethna, Simulation Product Development, Aspen Technology, 2500 CityWest Blvd, Houston, TX 77494.

The optimal operation of refinery crude pre-heat trains represents a significant challenge to the operations engineer. Frequent change in the crude feedstock and usual changes in the performance specifications create plant operation variations that are difficult to predict. Heat exchanger fouling compromises optimality and the usual implementation of ad hoc heat exchanger cleaning schedules compromises the optimal operation of the pre-heat train and reduces the overall profitability of the plant. Moreover, furnace fuel consumption is expensive and can be throughput limited if the fired heater capacity is reached. A three-step monitoring and performance improvement solution is presented that facilitates the involvement of the engineer in the daily pre-heat train operation and maintenance.

In the performance monitoring step, reconciled plant data is used to estimate the extent of fouling for the individual heat exchanger units. This first step also predicts the impact of pre-heat train fouling on the furnace inlet temperature. The subsequent decision support step evaluates the economic impact using pre-defined scenarios. In this second step, the effectiveness of cleaning and the use of bypasses, among other operating changes, are evaluated. Rapid assessment of the impact on Key Performance Indicators (KPIs) is conducted by using Aspen HX-Net Operations®. The final decision implementation step compares KPIs for the selection of the most practical and cost effective solution.

Using Compact Welded Heat Exchangers to Optimize Heat Recovery in Refineries

Chris Wajciechowski P.E., Alfa Laval Packinox, 5400 International Trade Drive, Richmond, VA 23231

In today's refinery business, there is an increasing emphasis on expanding capacity; while at the same time reducing energy consumption and maintaining a positive environmental profile. The use of compact welded plate heat exchangers (CPHEs) can enable refiners to meet or exceed their goals in this key agenda. The inherent flow properties of CPHEs make them a very efficient, compact alternative to traditional shell-and-tube heat exchangers. CPHEs make it possible to achieve increased throughput in a smaller space, while significantly increasing energy intensity and lowering heater emissions. Maintenance can also be reduced by limiting deposition fouling with a CPHE. In order to gain the maximum benefit from CPHE, the performance of our equipment needs to be integrated into the heart of the process design. Often times, a simple replacement or the use of conventional Pinch Technology falls short of successfully exploiting the gains of CPHE. Therefore, during a revamp or energy project, the CPHE needs to be laced into the process to give the maximum benefit. In a crude preheat train revamp, energy reductions of 25% or more are possible when CPHE are strategically used alongside existing S&T exchangers. This discussion will show examples of how refiners have integrated CPHE in specific unit operations and how their process energy intensity has improved as a result.

Web Page: www.alfalaval.com/ecoreJava/WebObjects/ecoreJava.woa/wa/showNode?siteNodeID=1850&contentID=-1&languageID=1

SUCCESSFUL REVAMPING OF YR FIXED BED REFORMER INTO CCR

Mohammed A. Balamesh, Saudi Aramco, Dhahran, Saudi Arabia
Rabea M. Al-Saggaf, Saudi Aramco, Yanbu, Saudi Arabia

Abstract

Saudi Arabian Oil Company (Saudi Aramco) Yanbu Refinery successfully converted the existing 38 MBD fixed bed Reformer to continuous catalyst regeneration (CCR) to produce 40 MBD platformate with 100 Research Octane number Clear (RONC) gasoline blending component. This conversion was a major capital project that improved the Reformer unit reliability and efficiency, by reducing the down time for catalyst regeneration. This unit conversion was the first revamped Reformer/CCR unit in Saudi Aramco. The latest UOP regeneration technology cycle max has been used in this conversion.

This paper will share the revamping conversion of Yanbu Refinery Reformer to CCR, the technology used, lessons learned and the gained benefits.

Introduction to Refinery

Yanbu Refinery (YR) is a hydro-skimming refinery located on the west coast of the Saudi Arabia (KSA). The refinery was inaugurated on 1983 with a capacity of 170 MBD to sustain the KSA local demand of petroleum products. Currently, the refinery process 235 MBD of Arabian light crude oil and produce about 295 MBD of petroleum products, including imports. Production has included fuel oil, low sulfur diesel, two grades of premium gasoline, LPG, Jet fuel, and butane fuel. YR process units consists of atmospheric crude distillation unit (CDU), LPG Merox, Heavy Straight Run Naphtha (HSRN) hydrotreater unit , Reformer & Continuous catalyst regeneration (CCR) unit, Light Straight Run Naphtha (LSRN) hydrotreater, Isomerization unit , Saturation Gas Concentration unit (SGCU), Amine treating unit and Diesel Hydrotreater (DHT). A block diagram of YR is shown figure (1).

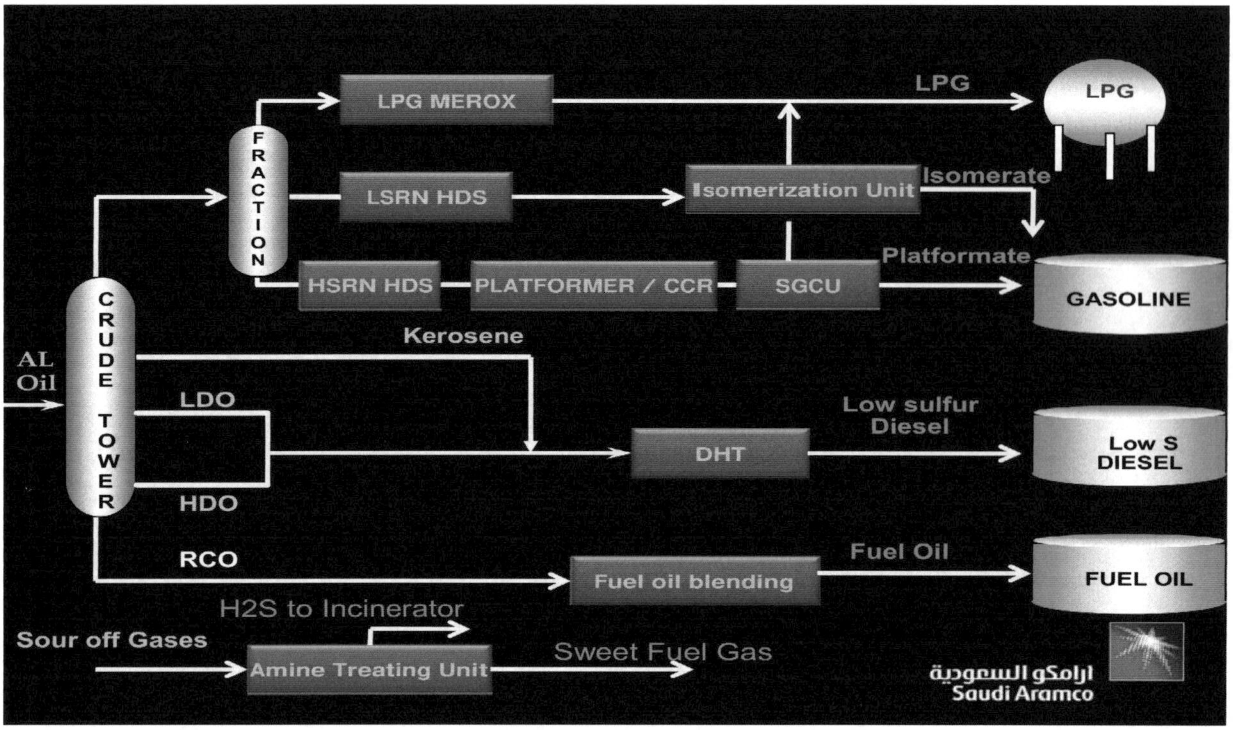

Figure (1) - Refinery configuration block diagram

Refinery Fixed Bed Reformer Unit

A Reformer unit is a refinery process that uses selected operating conditions and catalyst to produce high octane component for motor fuel blending. The feed to the Reformer unit is heavy naphtha feedstock which is hydrotreated to lower sulfur and nitrogen contents to maximum 0.5 ppm in the naphtha hydrotreater unit. The treated naphtha is combined with recycle gas and heated by the reactor effluent exchanger. The Reformer heaters then process the feedstock over the catalyst bed to produce a high octane number. The product is then sent to the gas concentration unit for final vapor pressure adjustments. The net gas hydrogen product from the reactors is fed to refinery users and excess gas to the fuel gas system.

The YR Reformer Unit was originally designed as a fixed bed unit to process 35 MBPD Heavy Straight Run Naphtha (HSRN) and to produce 26.25 MBPD platformate at 94 RON. In 1995, the unit throughput rate was increased from 35 to 38 MBD and operating severity of 96 RONC. The unit was shutdown every 6~8 months as part of a cycle to regenerate and restore the activity of the Platforming catalyst.

YR Reformer originally consists of three fixed bed reactors that utilize UOP Platforming UOP R-56. The unit reheat section has one charge heater and two interheaters to maintain the required reaction temperature at each catalyst bed of the reactors. In addition, a total of four vertical combined feed exchangers preheat the charge feed and cool the last reactor effluent product platformate. As well, two recycle gas compressors are used to recycle the required hydrogen for the Platforming reactions. Additionally, total of three make up gas compressors are used to make up the required H2 consumption in the refinery hydrotreating units. Also, the unit has the gas/liquid separation section consist from product separator to separate the rich gas from the product liquid platformate.

YR Reformer fixed bed process flow diagram is shown in figure (2).

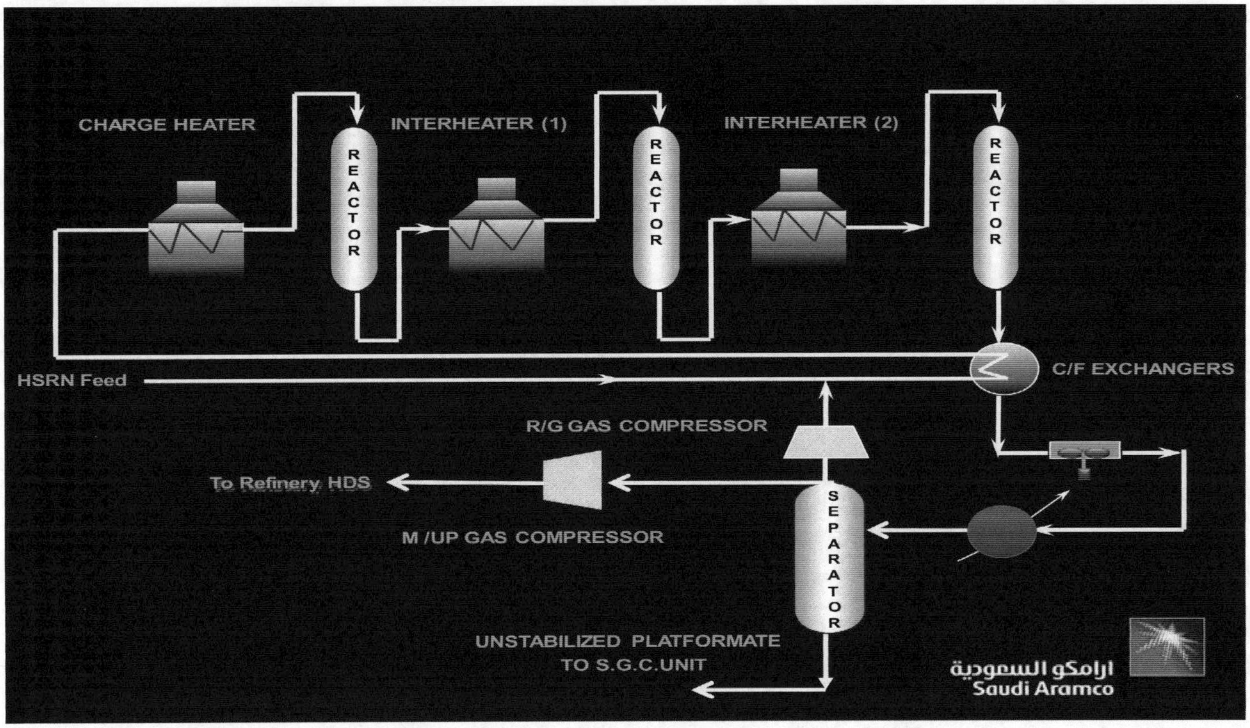

Figure (2) - Refinery Reformer Process flow diagram

Revamp objectives

In Oct. 2002, a capital project was approved to convert the old Fixed Bed Reformer to Continuous Catalyst Regeneration Reformer (CCR) unit. The project proposal been done by Foster Weller company while the detailed design, procurement and construction was awarded to Snamprgetti company.

The main objective of this revamp was to increase the volume of the refinery's unleaded gasoline pool to help meet Kingdom gasoline demands and reduce imports. The additional 95 RON gasoline production specification will be approximately 8,000 BPD and the annual revenue will be in the millions.

Also this will reduce the requirement for expensive purchased blending components such as MTBE and will allow for blending of higher volumes of LSRN and butane and other lower cost components into the gasoline pool. On the other hand, this revamp will increase the unit service factor, since downtime for regenerations will be eliminated.

These goals are met by the introduction of higher octane, higher volume production, made possible by the Reformer-CCR revamp.

The revamped will perform the same function as the unit it replaces, i.e., the reforming of heavy naphtha to high octane reformate. The revamped CCR Reformer will produce 100 RON Platformate at 75.5 % C5+yield, 230.3 nm^3/m^3 (H2/feed), and 40 MBD charge rate, rather than the current fixed bed Reformer production of 96 RON at 76 % yield and 38 MBD maximum charge rate. Another functional requirement is continuous catalyst regeneration, which will maintain RON, yield and charge rate capability without the need for regeneration downtime.

The conversions revamp strategy

After the completion of all construction and pre-commissioning activities, it was planed to shutdown the old fixed bed unit to do the required modifications to change the unit configuration to continuous catalyst regeneration. Also during the shutdown the regeneration of the old catalyst has been completed before starting the mechanical activities. The regeneration of the old catalyst was done to prepare the catalyst for dumping, screening and platinum recovery. The regeneration included only the carbon burn step without chloride injection which saved approximately 12 M$. The regeneration was completed successfully where the carbon level after carbon burn was 0.35 wt%. After completing the regeneration the old fixed catalyst R-56 was unloaded and screened for metal recovery. Then the old three reactors were demolished. Around 256 mechanical tie-ins have been completed during this shutdown which took 28 days to complete all the above mentioned work.

Refinery Reformer after Revamp

YR fixed bed Reformer unit has been revamped to CCR on June, 2006. The major changes for this conversion are the following.

► **Reforming Stacked Reactors**

The old series reforming fixed bed reactors has been replaced by three new Reforming stacked reactors. The important feature of the stacked reactors is the low pressure drop to facilitate the catalyst movement from the reactor side to the regeneration side. The design pressure drop is 0.24, 0.23 and 0.31 kg/cm2 for reactors 1, 2 and 3 respectively.

► Cycle Max CCR

The revamp incorporated a new UOP continuous catalyst regeneration (CCR) cycle max model for the first time in operation within Saudi Aramco. The CCR is designed to continuously regenerate the Platforming UOP-R234 at 100 % circulation rate or 2000 lb/hr (907 kg/hr) to maintain the catalyst selectivity and activity.

► New CCR Refrigeration Unit

A new CCR refrigeration package was installed as part of the revamp. The objective of this unit is to increase the purity feeding the CCR reduction zone, lock hopper and the regenerated catalyst lift line.

► New Waste Heat Boiler

As part of the revamp and to optimize the refinery energy savings, a new Waste Heat Boiler (WHB) was installed in the main duct of the Reformer heaters. The WHB utilize the waste energy of the Reformer heaters flue gas to generate 10.5 kg/cm2 (150 PSI) steam production at design rate of 35 tons/hr with future spare capacity to produce 60 ton/hr.

► Platformer Heaters Modification

A new modification was carried out in the Reformer heaters to maintain the required heat duty to perform the Platforming endothermic reactions. The existing old second and third interheaters were combined together and converted into one interheater to maintain the second Reformer stacked reactor reaction inlet temperature. In addition, a new additional heater was designed to maintain the reaction inlet temperature of the last third Reformer stacked reactor. The total Reformer heaters after this modification are four heaters including the charge heater.

► Net Gas System Modification

The net gas system was modified as part of the revamp. The existing old make up gas compressors were reused to supply the H2 to the CCR unit. Two new compressors were installed to supply the refinery hydrotreating units (NHT, LSRN Hydrotreater) and Isomerization unit with the required H2.

► AET LPG Recovery Unit

A new AET (Advanced Extraction Technology) unit was designed as part of the overall revamp of the Reformer unit. The objective of this unit is to maximize the LPG recovery from the Reformer net gas stream and increase the hydrogen purity of the net gas to meet the required high hydrogen purity for the new diesel hydrotreater unit

(DHT). The AET is designed to recover 98 % of LPG and increase the purity of the net gas from 80 mol% to 89 mol % for use in the DHT unit.

► Net Gas Chloride Treaters

Two new chloride treaters were installed on the makeup gas to the refinery hydrotreating units and the net gas feeding the new diesel hydrotreater (DHT) unit. The purpose of these chloride treaters is maintain low chloride level (max. 1.0 ppm) in the effluent hydrogen gas to guard the downstream unit users from corrosion related problem.

The YR Reformer reactors system revamp simplified process diagram is shown in figure (3). The new continuous catalyst regeneration interaction with the reactor system is shown in figure (4).

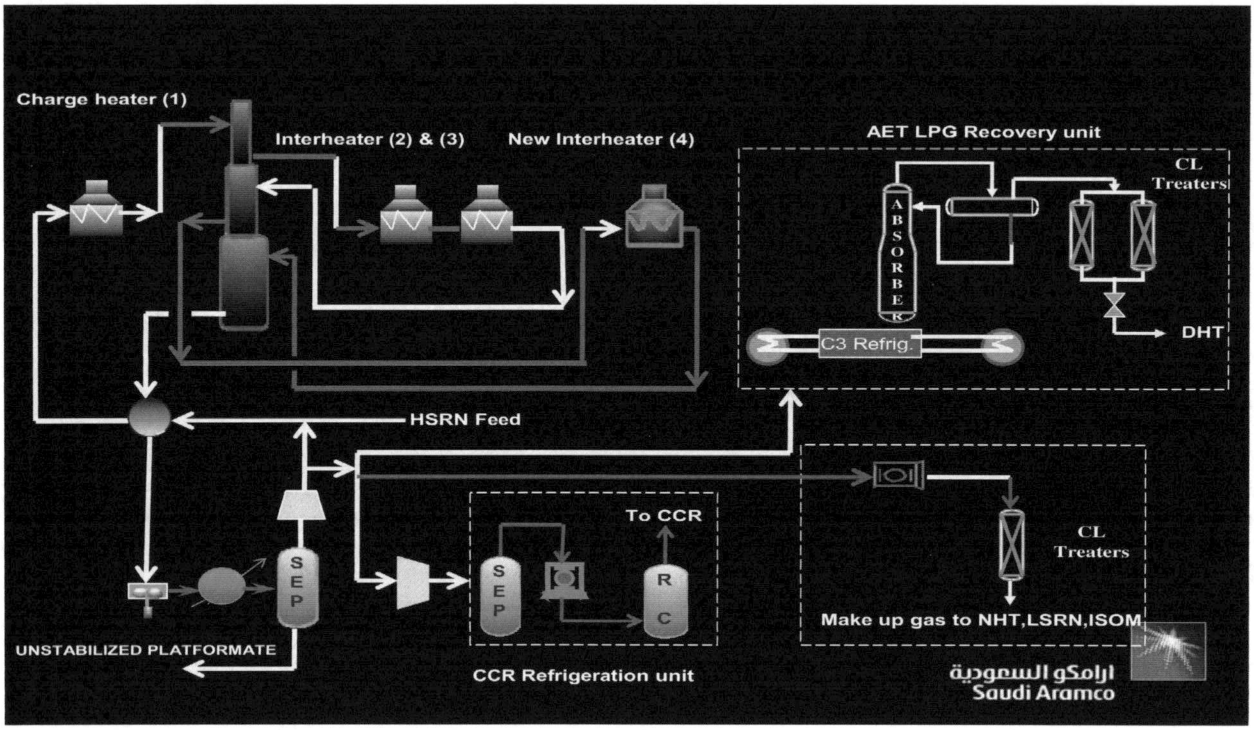

Figure (3) - Refinery Reformer after Revamp process flow diagram

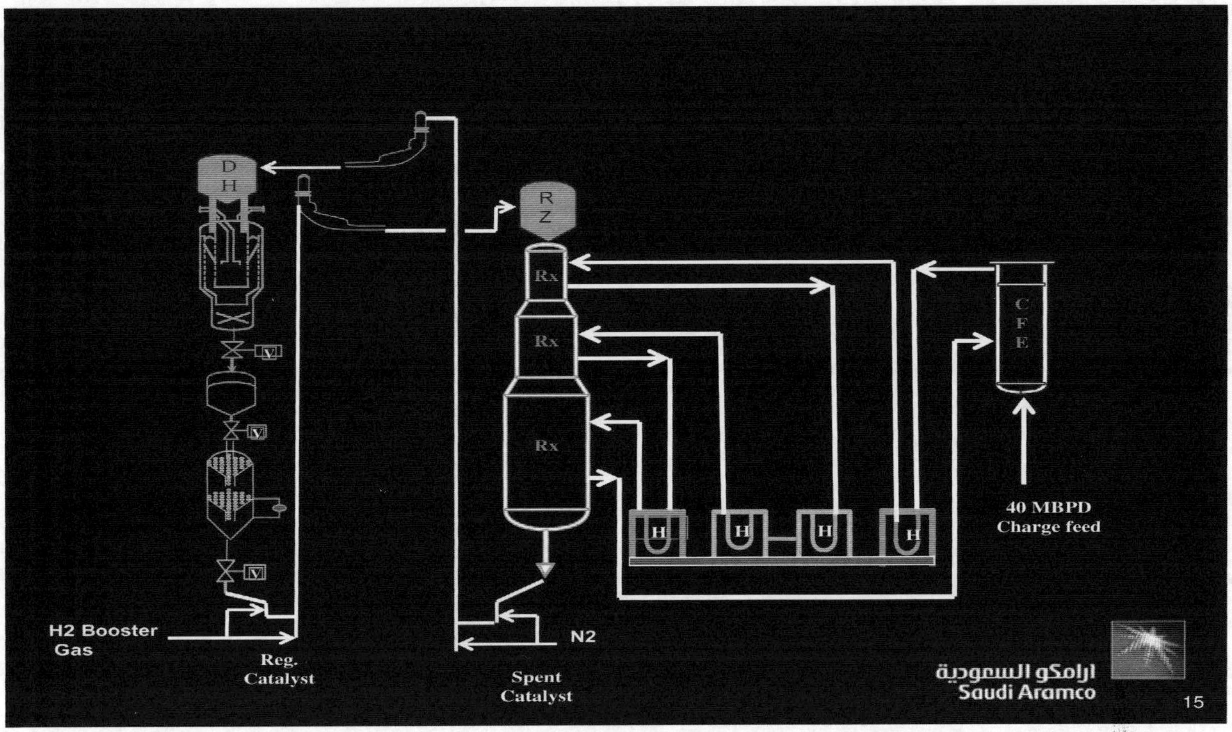

Figure (4) - Refinery Continuous Catalyst Regeneration CCR

Performance Comparison

The performance comparison between the Reformer before revamp and after conversion to CCR is shown in table (1).

Table (1) – performance comparison before and after revamp

Criteria	Before Revamp	Guarantee
Type	Fixed Bed	CCR
Capacity (MBD)	38	40
Severity (RONC)	96	100
Yield %	74.5	75.5
Gasoline production (MBD)	48	56
Steam Production (Tons/hr)	-	35

Lessons Learned

The objective of this part is to capture and present useful knowledge to help future teams capitalize on successes achieved and avoid problems encountered during execution of this revamping project. During the revamping, several challenges were experienced, but preparedness and excellent teamwork had effectively minimized their impact on the project.

The following are the major lessons learned from the experiences of this revamping project:

Implement Y-Team approach and culture for other projects

Project stakeholders include PMT, YR, Contractor, Loss Prevention, Inspection, and other parties. Traditionally, the various stakeholders had different interests, lack of common objectives, lack of transparency, and inadequate planning and communication. This often led to project delays. YR initiated the Y-Team concept for this project, integrating all stakeholders into one overall team

Benefits of the Y-Team concept included:

- Reduced project cost, because everybody worked for the same the Company
- Shared and achieved common objectives
- Prepared planning/coordination for common activities such as DCS, Instrumentation, Electrical, etc.
- Created an enabling environment for operations

1. Utilize Automated MPAS System to Manage Exception Item and Mechanical Completion Certificate Approval

Previous projects used a conventional manual hardcopy report system for tracking exception items. YR projects have successfully used an automated system (MPAS) to automatically log, track, and close exception items.

Benefits of the automated MPAS system include:

- Expedite exception item handling through easy-to-use system
- Improve transparency
- Project accountability achieved
- Support focusing on project completion and closure
- Better tracking of exception items
- Ability to generate variety of useful reports, such as sort by project initiator
- Save paper

- Save time

2. Ensure the consistencies in vendor data

During the commissioning, it was observed that one of the equipment nameplates recommend using an oil type material while the equipment document and manual referred to different material types. The team contacted the vendor, who confirmed that the manuals type is the correct one. As lessons learned we should make sure that the name plate and the manuals of the equipments are matching each other.

Consider inspecting equipment

One of the vessels failed during the start up. The investigation revealed that there was no requirement governing the internal inspection of vessels. It was learned that vessels interiors should be inspected by the vendor as well as the proponents to avoid unexpected challenges.

Conclusion

YR 38 MBD fixed bed Reformer has been revamped to 40 MBD Continuous Catalyst Regeneration Reformer with a 100 RON at 75.5 % reformat yield. This new project will increase the refinery unleaded gasoline capacity by 8 MBD and the unit down time will be improved. The CCR project construction was completed by June 2006. The preparedness and excellent teamwork effectively helped to overcome all challenges, resulting in a stable operation with 40 MBPD throughputs, 100 RON platformate octane and 100 % catalyst circulation.

Optimisation of Preventive Maintenance in the Process Industry

Qiying Yin, School of Chemical Engineering and Analytical Science, Centre for Process Integration, The University of Manchester, PO BOX 88, Manchester, United Kingdom

This paper presents a systematic methodology for optimising the Preventive Maintenance (PM) interval based on mathematical programming. Failure Modes and Effects Criticality Analysis (FMECA) is used to identify the critical components with critical failure modes in a process. The Weibull distribution model is applied to present the failure rate of the critical component and on this basis, a new methodology is developed to describe the relationship between reliability and PM interval. The optimisation is to minimize the process lost production cost in a defined period. A case study is presented regarding a LNG process with capacity 50t/hour. The critical equipment in the process is found by conducting FMECA. The optimal solution obtained is compared with the solution in practice. Different economic scenarios can impact the optimal PM solution when optimised. This demonstrates that it is crucial to understand the economics scenarios involved and to account properly for the relevant cost implications. It is important to consider reliability availability and maintainability study at the process conceptual stage design. The improvement in plant availability from choosing the right process configuration in first place can easily overcomes the additional capital investment.

Characterization of the Durability of Nafion Membranes and Nafion/ Inorganic Oxide Nanocomposite Membranes

Kenneth A. Mauritz, Mohammad Hassan, David Rhoades, and Robert B. Moore. School of Polymers and High Performance Materials, University of Southern Mississippi, 2609 West 4th, Hattiesburg, MS 39401

Selected results in the investigation of the chemical degradation of Nafion perfluorosulfonic acid membranes in the fuel cell environment will be presented. The topics will include a detailed explanation of the use of the powerful technique of modern broadband dielectric spectroscopy in monitoring the shift in the dynamics of Nafion macromolecular chains with degradation. In particular, long range main perfluorocarbon chain motions associated with the glass transition are revealed, the signature of which is the distribution of relaxation times (DRT). It will be explained how the DRT is related to the shift in Nafion molecular weight distribution with degradation in peroxide via the Fenton's reagant test. Also discussed will be the concept and rationale of the use of Nafion membranes that are inorganically modified with membrane - in situ - grown nanoscopic fillers that render the membrane more hydrophilic and possess greater hydration at higher temperatures.

Microstructural Characterization of Life-Limiting Membrane-Electrode-Assembly Materials Degradation Phenomena

Karren L. More and Kimberly S. Reeves. Metals and Ceramics Division, Oak Ridge National Laboratory, 1 Bethel Valley Rd., PO Box 2008, Oak Ridge, TN 37831

Advanced electron microscopy coupled with a unique ultramicrotomy specimen preparation technique are being used to characterize the network of microstructural features within PEM fuel cell membrane electrode assemblies (MEAs) and to determine the roles of the material constituents contributing to performance loss during PEM fuel cell operation. Atomic-level structural and compositional analyses have been conducted on intact MEAs before and after life-testing to identify the degradation of the individual constituents comprising the MEA, including: (1) electrocatalyst (2) carbon catalyst support (3) proton conducting membrane (4) re-cast ionomer. In the work to be presented, Pt- and Pt-alloy-based 3-layer MEAs were characterized primarily by transmission electron microscopy (TEM). The MEAs were characterized in the fresh (as-processed) condition and then subjected to an extensive matrix of aging conditions to evaluate the effects of temperature, humidity, and potential cycling on the MEA constituent stability. Degradation occurred by a combination of catalyst particle coarsening, localized carbon corrosion, Pt dissolution and migration, and nm-scale membrane changes, and these observations will be described in detail.

Membrane Degradation of Proton Exchange Membrane Devices Operating in an Oxygen Deprived Environment

Ludwig Lipp[1], Pinakin Patel[1], **Trent Molter**[2], and Peter Menard[2]. (1) FuelCell Energy, Inc., Danbury, CT 06813, (2) Connecticut Global Fuel Cell Center, University of Connecticut, Storrs, CT 06269

Cation exchange membranes based on perfluoroionomers have been used extensively in fuel cells, electrolyzers and other devices for over 30 years. Frequently, these membranes are called upon to serve two important functions within an electrochemical cell. First they provide a barrier that effects a positive separation of reactants and/or products. Second, these membranes provide selective ionic conduction between the anode and cathode chambers of the cell. In this manner they help to complete the electrochemical circuit between the two half-reactions. In some cell designs, these membranes are also utilized as the central element in a gasket seal design, thereby preventing fluids leakage from the electrochemical cell to the external environment. Compromise of any of these important functions can lead to premature cell failure; limiting the overall reliability of the PEM system.

Today's service requirements for PEM electrochemical devices for stationary and transportation applications require the demonstration of thousands of hours of invariant performance under the challenging operating conditions. Extensive accelerated laboratory and in-service testing of PEM fuel cells designed for these applications has shown some form of degradation, manifesting itself in two modes. The first is a gradual unrecoverable performance loss often characterized by continual voltage decay at a specified current density. The second is a sudden failure of cell or stack associated with gas crossover through pinholes and tears or fissures formed in the polymer electrolyte membrane. Some work has been done to study each of these modes, however a complete picture regarding the mechanism of each of these degradation modes is not available.

While many researchers have studied the durability of PEM fuel cells, much is still not known about degradation modes and the effects of damage accumulation on operating cells. Extensive research has, however, been conducted to elucidate the chemical pathways for membrane decomposition. The detection of the decomposition products, such as hydrogen fluoride and other polymer chain fragments in the condensed effluent water provides a strong indicator of the health of perfluoroionomer membranes. Through these and other studies, researchers have indicated that most degradation mechanisms rely on the presence and participation of oxygen to either initiate or to propagate the failure mode.

But what is not well understood is what happens when the cell is removed completely from the oxygen environment. This paper explores the degradation characteristics of electrochemical cells based on PEM technology that are operated in an oxygen deficient environment. In this regard, degradation of PEM cells configured to operate in the hydrogen pump mode was characterized. Stack lifetime performance data are reported along with diagnostic studies of the condensate effluent and product gas quality from a 25-cell prototype stack. Performance results are compared to PEM fuel cells that use air as the oxidant.

Interfacial Durability Issues for Hydrocarbon Membranes in Pemfcs

Melinda Hill, Yu Seung Kim, and Bryan Pivoar. Los Alamos National Laboratory, P.O. Box 1663, MS D429, Los Alamos, NM 87545

Sulfonated poly(arylene ether)s exhibit comparable PEMFC performance to Nafion without the high cost associated with the perfluorinated membranes. However, the long-term performance of traditional hydrocarbon membranes is poor when paired with standard Nafion-based electrodes, due to inadequate interfacial compatibility which can lead to delamination during swelling/deswelling. This may be due to the fact that hydrocarbon membranes with sufficient conductivity usually have high water uptake. Incorporation of polar groups and partial fluorination can change the important membrane properties, including proton conductivity and water uptake. Interfacial resistance and long-term fuel cell performance of several hydrocarbon copolymers will be discussed.

Analysis of the Physical and Chemical Stability of Cast Nafion® Membranes with Respect to Methanol Solubility and Radical Attack

Shawn J. Osborn, Brady W. Pitts, Kenneth M. Mauritz, and Robert B. Moore. School of Polymers and High Performance Materials, University of Southern Mississippi, 2609 West 4th street, Hattiesburg, MS 39401

We have recently found that the new generation cast Nafion® membranes are susceptible to disintegration/dissolution when subjected to boiling methanol. In order to mitigate this undesirable behavior, we have begun a series of investigations aimed at improving the morphology-property relationships in these cast materials. The purpose of this study is to investigate the effects of thermal annealing and solution-processing parameters on the solubility of new generation cast Nafion®. In this work, it is shown that the percent solubility decreases upon thermally annealing the film in the acid form above 175°C or solution-processing a film around 180°C from a high boiling point solvent. With respect to chemical stability, we have begun studies aimed at examining the effects of Fenton's Reagent on the resistance to radical attack of new generation cast Nafion®. Changes in the 19F solid-state NMR spectra of new generation Nafion® before and after chemical degradation via Fenton's Reagent predicts a rather random attack by •OH and •OOH radicals with slight preference for the side chain. These spectroscopic data are in agreement with measurements of the ion exchange capacity (IEC) and equivalent weight (EW) of the polymer films before and after chemical degradation.

Sulfonated Perfluorocyclobutyl Polymers and Fluorocarbon Functionalized Carbon for Pem Membranes

Amit Y. Sankhe[1], Raul S. Hernandez[1], Mark W. Perpall[1], Chris M. Topping[1], S. Suresh[1], Jianyong Jin[2], Earl H. Wagener[2], Stephen Creager[1], and Dennis W. Smith[1]. (1) Clemson University, Department of Chemistry and Center for Optical Materials Science and Engeering Technologies, Clemson, SC 29634, (2) Tetramer Technologies LLC, Pendleton, SC 29670

Novel ways of using perfluorocyclobutyl (PFCB) based polymer systems as proton-exchange membranes will be discussed. PEMs based on direct sulfonation of bisphenyl-PFCB polymer derived from biphenyl trifluorovinylether (BPVE) monomer have been investigated. Direct sulfonation of the PFCB polymer provide an efficient way to control the degree of sulfonation. The sulfonated polymer also formed stable free standing films that were solution cast from DMAc. In addition, we have also investigated a new class of sulfonated poly(arylene ether sulfone) (PAES) containing the perfluorocyclobutyl (PFCB) unit, prepared by the polycondensation of a unique bis-phenol (Bisphenol–T) with two dichlorodiphenylsulfones (DCDPS) under nucleophilic substitution conditions. The incorporation of the sulfonated repeat unit into the copolymers was determined by NMR and FTIR. The resulting polymers show good solubility in polar organic solvents such as DMSO, DMAc and form tough, flexible films by solution casting. The polymers exhibit glass transition temperature near to 130 °C and catastrophic weight loss in nitrogen at 350-450 °C. We have also developed a number of porous carbons from polymeric precursors. Both traditional resorcinol-formaldehyde resin and BODA nanomolding chemistries were utilized to yield porous carbons of different pore sizes, for potential use as a self-supporting solid electrode with reliable electrical conductivity. Decoration of the carbon electrode surface with compatibilizing fluoropolymer and fluorinated acid sites, through aromatic diazonium salts will be investigated in this work.

Broadband Dielectric Spectroscopy Characterization of Chemical Degradation in Nafion®/Silicate Nanocomposite Membranes

Mohammad K. Hassan, Robert B. Moore, and Kenneth A. Mauritz. School of Polymers and High Performance Materials, The University of Southern Mississippi, P.O. Box 10076, 118 College Drive, Hattiesburg, MS 39406-0001

Nafion/silicate nanocomposite membranes were generated by means of in situ sol-gel processes catalyzed by dibutyltin dilaurate. The results of the use of broadband dielectric spectroscopy to examine the molecular motions in such hybrid membranes that were degraded in Fenton's reagent at room temperature will be presented. Spectral shifts in the β-relaxation (glass transition) temperature-frequency region for untreated samples that were caused by degradation will be discussed. Relaxation parameters extracted from fitting experimental data to the Havriliak-Negami and Vogel-Fulcher-Tammann-Hesse equations quantitatively characterize chemical degradation. These parameters reflect long range perfluorocarbon chain dynamics, local molecular environment and molecular weight distribution as affected by degradation. The results of other dynamic characterization methods, such as dynamic mechanical analysis, stress relaxation and creep will be presented and correlated to the dielectric spectroscopic results.

Transport Properties of Supercritical Fluid Processed Proton Exchange Membranes: Methanol Permeability and Proton Conductivity

David Suleiman[1], Juan C. Pulido[2], and Alba Y. Marrero[1]. (1) Chemical Engineering, University of Puerto Rico, Department of Chemical Engineering, Mayaguez, PR 00681, (2) Chemical Engineering, University of Puerto Rico - Mayaguez, Mayagüez, PR 00681

Proton exchange membranes (PEMs), commonly used in direct methanol fuel cells (DMFC), are typically limited by either high methanol permeability (also known as the cross-over limitation) or low proton conductivity. These membranes have phase segregated rubbery, glassy and ionic domains, which make processing challenging and limited by phase equilibria considerations. This study proposes to use supercritical fluids (SCFs) processing as an alternative, since the gas-like mass-transport properties of SCFs could allow for better penetration into the membranes and the use of co-solvents could also change the morphology of the PEMs, fine-tuning the physical and transport properties (e.g., fuel permeability and proton conductivity) needed for effective DMFC performance. This investigation focused on the effect of SCF processing on the physical and transport properties of tri-block copolymer ionomers commonly used as proton exchange membranes (PEMs). The PEMs selected included: Nafion®, sulfonated poly-(styrene-isobutylene-styrene) (SIBS) and sulfonated poly-(styrene-ethylene-ran-butylene-styrene) (SEBS). Processing was performed at 200 bars and 40 ºC using SCF CO2 with and without different co-solvents. A variety of co-solvents were used including: acetic acid, acetone, acetonitrile, cyclohexanone, ethanol, isopropanol, methanol, methylene chloride, and tetrahydrofuran. They were selected based on their size, polarity, and chemical nature (e.g., protic or aprotic). Measurements of methanol permeabilities were made using an FT-IR technique, monitoring the methanol concentration with time through the membrane. Proton conductivity was measured using impedance spectroscopy. Measured permeabilities were of the order of 10-8 cm2/s, two orders of magnitude lower than unprocessed membranes. However, changes in proton conductivity were short of the required level (0.1 S/cm) and very sensitive to the processing orientation. These results coupled with thermogravimetric, swelling, and dynamic mechanical analyses have provided a better understanding of the structure property relation of these complex three-dimensional phase-segregated membranes.

Sulfonated Block Copolymer Membranes

Dale L. Handlin and Scott R. Trenor. Kraton Polymers, 3333 Hwy 6 S., Houston, TX 77082

Numerous researchers have studied sulfonated styrenic block copolymers as semi-permeable membranes for various applications. Traditionally, the styrenic endblocks of triblock copolymers have been sulfonated, as was first practiced by Kraton Polymers as far back as the late 1960's. Under this process, once the polymer is hydrated, water plasticizes the endblocks which results in a significant degradation in the mechanical properties. Recently designed sulfonated block copolymer structures have been obtained that have continuous ionic domains and retain strength when fully hydrated. Utilizing 40+ years of experience in block copolymer structure and morphology, Kraton Polymers is now able to tailor the properties of the sulfonated polymer membranes via numerous methodologies including polymer structure, placement of sulfonic acid functionality, and membrane formation. The ability to tailor membrane properties provides the ability to manage the morphology of the membrane and decouple the degree of sulfonation and water swelling, as well as water swelling and transport properties. The initial characterization and the ability to manipulate the mechanical and transport properties of these polymers will be discussed.

AUTHOR INDEX

AUTHOR INDEX

AUTHOR INDEX

American Institute of Chemical Engineers
3 Park Avenue
New York, NY 10016-5991

ISBN: 978-1-60423-358-2